Tailings Dam Failure

Robert C. Lo

Tailings Dam Failure

Socio-Economic-Environmental Impacts

Springer

Robert C. Lo
Vancouver, BC, Canada

ISBN 978-3-030-87737-8 ISBN 978-3-030-87738-5 (eBook)
https://doi.org/10.1007/978-3-030-87738-5

This Springer imprint is published by the registered company Springer Nature Switzerland AG
The registered company address is: Gewerbestrasse 11, 6330 Cham, Switzerland

Dedication to my late father, Min Jee Lo, and late sister, Joanne C. Lo
Who had taught me many life lessons including a crucial one i.e.
"Strive to recognize fatal flaws in your pursuits at an earliest stage to prevent the consequences of their eventual failure."

Foreword

It is frequently said that all the commodities that we use in a modern society are formed from materials that are either grown by farmers or dug up by the extractive industries. Indeed, our fundamental reliance on those extractive industries, beyond perhaps hydrocarbons, is frequently forgotten or ignored. The scale of the extractive industries is vast—the 40 largest global companies alone have a combined revenue of over US$1 trillion per year.

Inevitably, the extractive industries generate byproducts. Some are very much in the news—greenhouse gases, in particular, have become a topic of true global significance. But mine waste is also very important, but receives far less attention.

In October 1966, the failure of a coal mine waste heap at Aberfan in South Wales killed 144 people, the vast majority of whom were children. In the aftermath of the disaster, an official tribunal (a public inquiry) examined all aspects of this disaster, producing a report that was scathing in its criticism of the operator of the site, the National Coal Board. The judge in charge of the tribunal wrote the following with regard to the operation of the tip:

> *We found that many witnesses, not excluding those who were intelligent and anxious to assist us, had been oblivious of what lay before their eyes. It did not enter their consciousness. They were like moles being asked about the habits of birds.*
>
> *As we shall hereafter seek to make clear, our strong and unanimous view is that the Aberfan disaster could and should have been prevented.*

The tribunal set out recommendations to ensure that mine waste was managed properly, rooted in a new legal and regulatory framework. In parallel, the scientific investigations that underpinned the investigation were the birth of our understanding of static liquefaction of mine waste materials, a major mechanism of failure.

But, almost 60 years after the Aberfan disaster, mine waste tragedies continue to occur, most notably (but not exclusively) in tailings. Failures in tailings storage facilities (TSFs) have occurred with depressing regularity over the last six decades, costing lives; inflicting huge social and environmental damage; and costing billions in mitigation costs and regulatory penalties. Even though our knowledge of the potential mechanisms of failure, and the ways that these can be mitigated, has improved, the rate of failure has accelerated. There are now estimated to be up to 35,000 tailings storage facilities worldwide, storing over 500 billion cubic metres of waste. Every year, this volume increases by about 19 billion cubic metres.

In the context of tailings, there are two fascinating recommendations from the Aberfan tribunal (amongst many others):

> *All tips should be treated as potentially dangerous* and
>
> *Tips should be treated as civil engineering structures*

These are wise words, and they are as true of tailing storage facilities as they are of coal mine slag heaps. No other area of civil engineering would accept the observed failure rate of tailings storage facilities, given the extraordinary consequences of such events. Yet, somehow, TSFs have been largely immune from the scrutiny to which other structures are subject.

Thus, there is a need to improve the planning, management, operation, decommissioning and legacy of TSFs. This volume details marked failures in every one of these areas and in many cases across a range of them. Facilities have been poorly sited; the characteristics of materials involved have been poorly understood; facilities have been poorly designed—in some cases, good designs have been undone by modifications that have introduced weaknesses. Facilities have been poorly constructed. Frequently the plan to process and deposit the tailings has been deficient or has been changed on the fly. Key structures, such as drains, have been neglected. And in many cases, the TSFs have been abandoned in the landscape once mining has ceased, allowing deterioration with time that ultimately leads to failure.

Some progress has been made in recent years, with initiatives such as the Global Tailings Management Institute. This was established in early 2025 specifically to act as an independent body "to drive and assess the implementation of Global Industry Standard on Tailings Management". This initiative, which is a collaboration across sectors, has the potential to make a real difference to the ways in which tailings are managed.

However, realistically at least three major challenges will remain:

1. There is a legacy of active TSFs that were designed for different climatic conditions and/or with a failure to understand seismic hazard—in effect, these are facilities with inadequate designs or that have poor construction. Retrofitting those facilities will be exceptionally expensive and very arduous;
2. there are thousands of abandoned TSFs around the world that are now being neglected and that will be deteriorating with time. Even determining who is responsible for such facilities, let alone managing them, is very uncertain. In many cases, responsibility will fall on the taxpayer;
3. a huge volume of mining occurs in unlicenced or poorly regulated environments. Proper tailings' storage and management are very expensive, so it is almost inevitable that unsafe structures will be constructed.

Notwithstanding this, every case in which a potentially dangerous TSF is upgraded or improved, or in which operational management is improved, reduces the risk burden of TSFs, and potentially saves lives. Whilst we will not solve the problem in our lifetimes, we can and must make substantial progress.

This volume collates our contemporary understanding of TSFs, and systematically reviews many case studies. As such, it is an essential and invaluable contribution towards reducing the risk portfolio for TSFs.

London, UK

Prof. David Petley

Preface

My graduate study was carried out from 1964 to 1969 in Cambridge, Massachusetts, USA, majoring in civil—geotechnical engineering and minoring in engineering geology and water resource engineering. I received both master of science and Ph.D. degrees under late Professor Arthur Casagrande. Thereafter, I worked on Tarbela hydroelectric project in west Pakistan and civil and mining projects with two internationally renowned consulting firms in USA and Canada before retirement. My experience covers: project management/review and advanced research. I have been involved in major civil and mining projects in North and South America, Asia and Europe and have published widely in technical journals, at conferences and online. I also have been active in national and international committees on earthquake, and tailings dam engineering as well as have participated in post-natural-disaster investigations.

Purpose of Book

This monograph serves as a compendium in addressing the issue of tailings dam failure and to document relevant references about a selected set of case histories in a recent period from 1965 to the present.

This book is written for practitioners whose work or interest is related to tailings management as well as the general public. Thus, the intended readers would include designers, operators, managers, financiers, insurers and regulators of tailings facilities as well as the general public. The former group is either working for or within a mining company to develop an eco-sustainable tailings storage facility that is cost-effective yet maintains the quality and safety of the facility's surroundings. The latter group is either directly impacted by the performance of the facility or has the general interest to safeguard the public from any undesirable effects resulting from the existence of a tailings facility, particularly when it malfunctions and/or fails.

The modest aim of this book is to provide an easy-to-understand and neutral-perspective reference monograph to facilitate ongoing dialogues in the public as well as earnest discussions among all stakeholders in working together for achieving the goal of tailings dam safety.

Vancouver, Canada Robert C. Lo

Acknowledgements

The author would like to thank the following prominent engineers who had contributed to the development of sustainable tailings management practice in its critical growth phase and their guidance to the author to carry on this practice:

late Mr. Earle J. Klohn—who had introduced the author to the realm of good tailings management;
late Mr. Charles F. Ripley—who had offered important advice willingly;
late Mr. Mark T. Olsen—who had demonstrated sound practice in actual projects; and
late Mr. C.H. (Bob) Maartman—who had offered insightful project reviews.

The author is grateful for receiving the encouragement and critical comments from his colleague, Mr. Len Murray regarding publishing this book.

The author would also like to thank Ms. Ana Rosa Blue for her assistance in locating relevant references and Ms. Danielle Leung for her preparation of the final Word document of the manuscript.

The professional guidance and assistance of Ms. Margaret Deignan – Sr. editor, Mr. Ambrose Berkumans – associate, two reviewers of Springer Publisher and Mr. Vishnu M. G. of Straive are greatly appreciated.

Last but not least, the author would like to thank his wife, Lisa, and daughter, Grace, for their continual support to allow this book becoming a reality.

Introduction

A well-performed tailings dam serves to store an enormous amount of mine tailings, which are the fine-particle solid waste by-products produced from the modern mineral extraction process. However, the failure of such a large hydraulic structure would flood the downstream area with water and tailings mud mixed with potentially toxic elements. This nightmarish scenario is occurring with increasing frequency and disastrous consequences around the world. In response to this alarming trend, the International Commission on Large Dams (ICOLD), the organization established in 1928, issued a world declaration on dam safety on October 18, 2019, making it the highest priority of the commission. It further called upon Governments, Financial Institutions, and other Developers to make a similar political and financial commitment to safety improvements and risk reductions at all dams.

Book Organization

The monograph consists of the following three chapters:

- Chapter 1—Mine Tailings, Tailings Storage Facility and Tailings Dam;
- Chapter 2—Dam Failures, Emergency Preparedness and Response; and
- Chapter 3—Representative Case Histories and Lessons Learned.

Chapter 1 Mine Tailings, Tailings Storage Facility and Tailings Dam

Abstract Chapter 1 introduces tailings as a fine-grained mine waste by-product originated usually from the milling process. Tailings characteristics vary with the type of minerals involved as well as the flow sheets of the milling process. Both the physical and chemical characteristics of tailings influence the design of an optimal facility for its storage. Considerable savings can be achieved, when tailings material itself can be effectively used to construct tailings dam(s) required for tailings storage.

Tailings storage facilities can be grouped into following categories:

- Underwater disposal in natural water bodies—In the past, tailings have been directly discharged into natural water bodies such as streams, lakes, or oceans without worrying about the environmental ramifications of this practice. Currently, underwater disposal is still being practiced under special circumstances. However, its environmental-socio-economic impacts tend to be scrutinized more closely with a thorough investigation, trade-off consideration, and adjudication.
- Underground storage—The primary objective of using tailings as underground mine backfill is to provide a convenient and stable working floor and/or lateral support to underground workings. The backfill also mitigates the long-term ground subsidence problem in the mined-out area. If the backfill is below the long-term water table, prevention of oxidation of acid-generating tailings is another important consideration.
- Above-ground storage—Above-ground storage can be grouped into the following three types:
  - dry tailings facility (no peripheral tailings dyke);
  - thickened tailings facility (with low peripheral tailings dyke);
  - tailings impoundment (with peripheral tailings dam(s)).

Tailings dams of early vintage and those constructed in low seismic and especially dry climate regions are often constructed by a procedure involving a minimum of engineered-fill placement. This was achieved by building a low Starter Dyke and discharging tailings from the top of the dyke until the impoundment was nearly filled with tailings. Then a second small dyke was constructed over the deposited tailings, and the procedure was repeated so that the tailings dam rose gradually in the upstream direction, hence the name "upstream method" was coined.

A much safer dam was the downstream dam, designed and constructed as a conventional water-retention embankment dam. This type of dam was also much more costly. The third type was the centreline dam that can be designed and constructed taking advantage of the inherent characteristics of a tailings dam. Thus, it was substantially less costly than, but just as safe as, a downstream dam.

In terms of storage capacity, tailings impoundment provides by far largest volume needed in modern mines.

Keywords Tailings · Tailings Storage Facility · Tailings Impoundment · Tailings Dam

Chapter 2 Dam Failures, Emergency Preparedness and Response

Abstract Chapter 2 covers potential dam failure modes, failure impacts, and factors affecting dam safety. Dam failure modes include:

- slope instability;
- dam overtopping;
- earthquake damage;
- foundation instability;
- seepage/internal erosion; and
- failure of components of flood-control hydraulic structures.

Impacts of tailings dam incident or failure on the affected communities include:

- loss of life and/or injury;
- loss of property and infrastructure; and
- loss of livelihood, which could be caused by loss of employment due to mine shut down and/or flood damage and environmental pollution.

Two main potential factors that contribute to dam mishaps are natural and human hazards.

Besides running a normal and cost-effective tailings dam operation, the mine operator needs to keep vigilance in anticipating which component of the dam system could go wrong? What are the potential consequences of such a mishap? How to respond to the mishap and minimize its impacts? Thus, it is critical to have a well-thought-out emergency response plan that details concrete actions to be taken by mining operator personnel at all levels. The plan also sets out well-defined internal and external emergency communication protocols.

An important document that is crucial in minimizing fatality suffered by downstream communities is the evacuation order. Concurrent to taking immediate remedial measures to prevent a dam breach, the mine operator should not overlook its critical responsibility of assessing the site situation realistically and conservatively to allow sufficient time for an orderly evacuation of the downstream communities to escape potential drowning by water or burying by tailings mud. Close liaison should be maintained with the government emergency agency so that the authority would have adequate time to initiate and execute the evacuation procedures to avoid loss of life. While unwarranted false alarms are to be avoided, appropriate conservatism is needed to make this all-important emergency call timely.

Keywords Dam Failure Modes · Emergency Preparedness · Emergency Response · Evacuation Order

Chapter 3 Representative Case Histories and Lessons Learned

Abstract Case histories of dam incidents and failures contain invaluable sources of practical and insightful information to guide practitioners in conducting the normal operation, in establishing safety-conscious practice and in enforcing rules and regulations to safeguard people's life and property, and in keeping the environment free from pollution. The selection of cases presented in Chap. 3 is based on databases available to the author, including: (1) collections by dam-building institutions such as International Commissions of Large Dams (ICOLD) and its national commissions; (2) technical papers; (3) forensic reports tasked by government agencies and/or dam owners; and (4) news articles. These cases include common dam failure modes: static slope instability; dam overtopping; seismic slope instability; foundation instability and seepage induced (piping) internal erosion.

Collected in this monograph are twenty-one select cases that occurred in developed as well as developing countries scattered around the world during different stages in the life span of respective tailings dams. These cases are presented in chronological order from 1965 to 2021. Some of these cases played an important role in influencing the evolution of tailings dam practice. These influences are highlighted in relevant cases.

Keywords Case History · Lesson Learned

General References

Books, Papers and Conference Proceedings

Tailings and Mine Waste 1978–2025.—ongoing conference series.

Vick, S. G. (1990). *Planning, design, and analysis of tailings dams*. BiTech Publishers Ltd.

- A classic textbook dedicated to tailings management. It introduces the subject clearly and thoroughly to readers and serves as an excellent reference book for whoever getting involved with the subject.

Marcus, J. J. (Ed.). (1997). *Mining environmental handbook—Effects of mining on the environment and American environmental controls on mining*. Imperial College Press.

- A comprehensive handbook about the effects of mining on the environment and the response of the United States to exercise environmental control on mining. Tailings management is only one component of this handbook that covers other mining issues besides tailings.

Szymanski, M. B. (1999). *Evaluation of safety of tailings dams*. BiTech Publishers Ltd.

- A comprehensive reference book by an experienced practicing consultant on tailings dam safety evaluation.

Paste and Thickened Tailings 1999–2025.—ongoing conference series.

Meggyes, T., Roehl, K. E., & Dixon-Hardy, D. (Ed.). (2005). *Tailings management facilities*. EPP Publication.

- An extensive state-of-the-art review of tailings management facilities based on European case studies concerning the worldwide experience. The review was carried out over a period from 2002 to 2005 for the Sustainable Improvement in Safety of Tailings Facilities (TAILSAFE) project.

Blight, G. (2010). *Geotechnical engineering for mine waste storage facilities*. CRC Press, Taylor & Francis Group.

- A comprehensive, recent textbook covers geotechnical aspects of mine waste storage facilities, containing practical information from design, construction/ operation to closure. The book is based on extensive consulting, research, and teaching work by Prof. Blight over the last four decades. It is a practical reference book for readers engaged in tailings engineering work.

Kossoff, D., et al. (2014). Mine tailings dams: Characteristics, failure, environmental impacts, and remediation. *Applied Geochemistry*, 51, 229–245.

- The review paper provides a general background of mine tailings and its retaining dams before focuses on causes and environmental and economic impacts of tailings dam failures and options for remediation measures.

Websites

Caldwell, J. (2010). *Geotechnical engineering for mine geowaste facilities*. EduMine Online Course.

Caldwell, J. (2012). *Tailings facility design, operation and closure*. EduMine Online Course.

http://www.tailings.info/index.htm Engels, Jon

- A website developed by Dr. Jon Engels based on his Ph.D. thesis (2006) at the University of Leeds. The website has been available to the mining industry and public since 2002. It covers information about tailings handling and storage.
- Its affiliated website, http://www.tailpro.com/resource, provides information on specialists working on diverse aspects related to tailings management practices.

https://www.tailingsnews.com.au/ Tailings News Alerts

– A newsletter website that covers news and information on tailings dams and engineered tailings storage facilities.

Contents

Abbreviations

ASCE	American Society of Civil Engineers
ICMM	International Council on Mining and Metals
ICOLD	International Commission on Large Dams
TNA	Tailings News Alerts
UNEP	United Nations Environment Programme

List of Figures

List of Tables

Chapter 1
Mine Tailings, Tailings Storage Facility and Tailings Dam

Abstract This chapter introduces tailings as fine-grained waste by-products originated usually from the milling process. Tailings characteristics vary with the type of minerals involved as well as the flow sheets of the milling process. Both the physical and chemical characteristics of tailings influence the selection and detail design of an optimal tailings storage facility. As the amount of tailings generated in a modern mine is enormous, therefore whenever appropriate, the use of tailings material itself to construct its storage facility makes good economical sense.

Keywords Tailings · Tailings storage facility · Tailings impoundment · Tailings dam

1.1 Introduction

There are three main options of tailings storage facilities:

- underwater disposal in natural water bodies;
- underground storage; and
- above-ground storage—above-ground storage can be grouped into the following three types:
  - dry tailings facility (no peripheral tailings dyke);
  - thickened tailings facility (with low peripheral tailings dyke); and
  - tailings impoundment [with peripheral tailings dam(s) on one or all sides, depending on site topography, to enclose the tailings impoundment].

This monograph mainly addresses the above-ground tailings impoundments, as they have large storage capacities, and any mishaps of them could have serious consequences.

According to their construction methods, tailings dams are grouped into three types:

R. C. Lo, *Tailings Dam Failure*, https://doi.org/10.1007/978-3-030-87738-5_1

- Upstream dam—the dam is raised from a starter dam to its ultimate height in the upstream direction;
- Downstream dam—the dam is raised in the downstream direction to its final height; and
- Centreline dam—the dam is raised in a zig-zag fashion, maintaining the same centreline from the starter to the ultimate dam.

The above description of tailings dam types according to the direction of dam raise towards: upstream, downstream or upward (centreline) is easy to remember. The upstream dam is historically earliest tailings dam, least costly and prone to failure. It remains to be used in the third decade of the twenty-first century probably is due to the following two reasons: (1) cost advantage, and (2) its selection by the mine operator often with little input from those with a great deal at risk, if and when the dam fails. The downstream dam is direct application of water retention dam for tailings storage without taking advantage of the idiosyncrasies of tailings dam. Thus, it tends to be most costly. The centreline dam does incorporate special features of tailings dam function and construction in its design and result in an optimal dam, robust and economical.

1.2 Mine Tailings

Tailings, sometimes referred to as fine-grained waste solid particles, are generated in the process of winning minerals from the ore. In other contexts, tailings are also referred to as both the fine-grained waste solid particles and the liquid effluent being discharged and stored together with the solids. Since tailings are essentially waste by-product produced in the milling process, their handling and disposal have been relegated historically to a relatively low status in the overall mining scheme. As tailings was considered merely as an inevitable item associated with the cost of mineral extraction, little consideration had been given to tailings other than minimizing its associated handling cost. However, the following four important trends have gradually rendered the above offhand way in tailings management out-of-date, and incompatible with modern sustainable mining practice that safeguarding public safety and the environment.

1. Past "laissez-faire" approach has led to cases of serious degradation of the environment including surface and ground water, air and soil surrounding a mining region because of contamination by pollutants due to geo-chemical processes involving acid drainage and/or metal leaching;
2. Incidents or failures of tailings facilities have too often resulted in fatalities and/or injury to people, loss of property and infrastructure and major environmental contamination. In the worst cases, the sudden release of tailings and supernatant water from the storage facility resulted in flood and/or debris flow that inundating the downstream area;

3. Modern surface mining open-pit operation (see Fig. 1.1) employs mining equipment of large capacity and mills of high throughput (see Fig. 1.2). While this development permits profitable extraction of even low-grade ore such as porphyry ore because of "economy of scale", it generates enormous quantities of tailings that call for safe major storage facilities (see Fig. 1.3); and
4. The social impact of the facility is often quite severe on downstream communities, who have to cope with major life-change or even life-threatening transformation related to the development and/or incident and failure of the tailings facility, while reaping only limited benefits.

Tailings slurry density (or pulp density) is the ratio expressed as percentage of the weight of solid divided by the combined weights of both solid and liquid in a tailings slurry. Slurry density is an important parameter that affects the behaviour of tailings slurry while being transported through pipelines or flumes. Moreover, upon

Fig. 1.1 Valley open pit (Highland Valley Copper)

Fig. 1.2 Grinding mills (left) and flotation cells (right) (Highland Valley Copper)

Fig. 1.3 L-L tailings dam—downstream retention dam of valley tailings impoundment (Highland Valley Copper)

deposition the segregation behaviour of tailings particles is also governed by the slurry density. As the density exceeds a certain threshold value, its deposited particle distribution stays the same without segregation.

In general, to produce concentrates from ore bodies, blasted rocks containing ores first need to go through crushers and grinding mills to reduce their size to fine-grained particles amenable to the floatation process that separates the heavier concentrates from the lighter tailings waste with the aid of chemical reagents. Once the valuable metals or materials are removed, the tailings stream is often subjected to a dewatering process to reduce the amount of water to be transported with the solid tailings in the form of slurry to the storage facility.

1.2.1 Tailings Characteristics

Kossoff et al. (2014) provided a recent review of mine tailings dams that covered tailings characteristics, tailings dam failure, its environmental impacts and remediation. The following factors exert strong influences on the physical and chemical characteristics of tailings:

- mineral composition and milling process determine the inherent tailings properties. Tailings involving substantial clay-particle sizes usually settles and consolidates slowly, posing difficulty in its storage;
- tailings is either discharged from large-diameter slurry transportation pipelines or spigotted from small-dimeter distribution pipelines at locations along the tailings dam crest and/or the perimeter of the tailings impoundment;

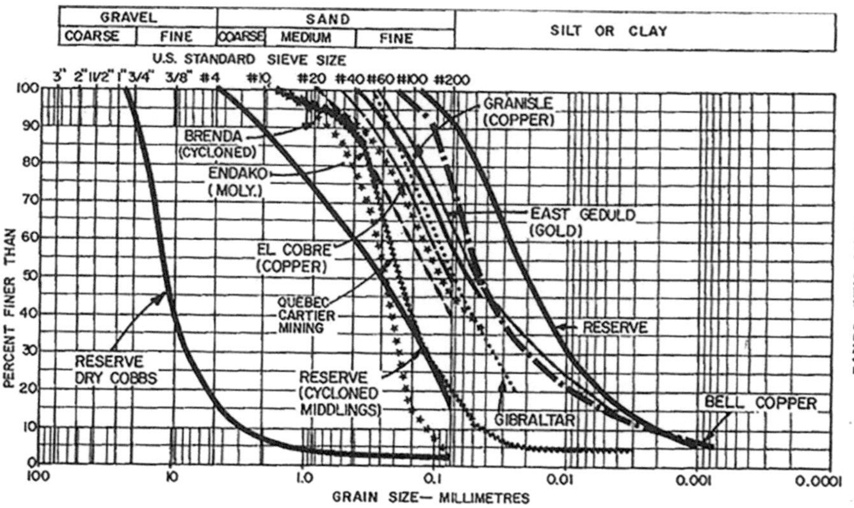

Fig. 1.4 Typical tailings grain size curves (after Klohn, 1982)

- as tailings flows from pipe discharge along tailings beach at ordinary slurry density (~ 30%+ solids), it goes through the sorting process involving particle saltation and rolling similar to that occurring in braided streams. This causes the coarser sand size particles to preferentially deposit near the discharge point. Further away from the discharge, tailings slope flattens out and the fines content increases as coarser particles are progressively removed by settling. Depending on the actual tailings deposition pattern, tailings particle-size distribution varies both horizontally and vertically along the beach. Tailings generally forms gentle beach slope varying from 0.5 to 2% depending on tailings particle size distribution (see Fig. 1.4) and slurry discharge water content and velocity;
- for high-density slurry (usually at 60% solids or higher) with substantial fines content, tailings deposit on the beach remains non-segregated and maintains relatively uniform, original particle-size distribution at the location of deposition (see Fig. 1.5). The threshold slurry density at which tailings exhibit such non-segregation property can vary significantly, depending on both the fines content and the colloidal chemistry associated with such fines (see Fig. 1.6—slurry density vs. segregation characteristics).

In cold climate region, ice lenses often are formed in the winter and later get buried. These ice lenses tend to impede vertical downward seepage, until they are gradually thawed out. Sometimes the thaw occurs long after cessation of tailings deposition, with sinkholes forming on the tailings surface as voids created by thawing being filled in by tailings above. In addition, surficial tailings in cold region is subject to freezing-induced consolidation. In arid climate region, tailings deposit is subject to air-drying and development of suction (negative pore pressure), which is conducive to tailings consolidation.

Fig. 1.5 Example of thickened tailings (thickened tailings images website)

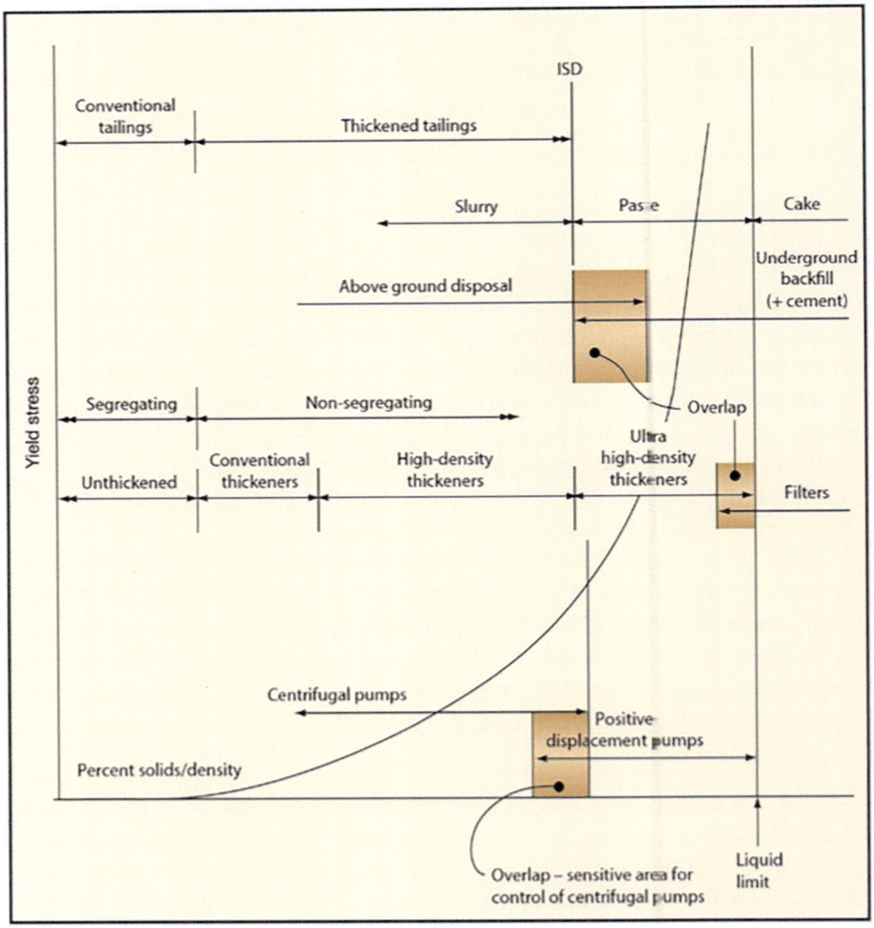

Fig. 1.6 Particle segregation versus slurry density (after Jewell & Fourie, 2022). *Note ISD* initial settled density

- Settling and consolidation—Tailings on the beach or in the supernatant pond first settles out through the sedimentation process, then consolidates under the self weight of additional tailings accumulated above; and after deposition, tailings undergoes physical and chemical weathering process. Chemical weathering could contribute to the problem of acid generation or metal leaching in initially harmless tailings. Chemical alteration can also involve cementation process that increasing bonding at tailings particle contacts.

1.2.2 Tailings as Dam Construction Material

The function of a tailings impoundment is to safely store tailings generated from the mineral extraction process. Thus, the opportunity to utilize tailings material itself to construct retaining structures for tailings storage has been pursued vigorously. The advantage is two folds: (1) the only cost associated with using tailings as dam-fill material is the incremental cost related to the engineering design and tailings operational modification that are required to make tailings utilization feasible; and (2) any tailings that are used for dam construction will require no further cost for its storage. Unfortunately, when insufficient consideration is given to the required adaptation details, inherent weakness could be built into the dam that may prove to be difficult to rectify later. Thus, caution is needed to fully take advantage of this cost-saving opportunity without incurring shortcomings in the as-built dam.

Tailings Dam Construction Material

As further described in Sect. 1.4, tailings dams can be classified into three types according to the method of construction: downstream, centerline and upstream method. In general, tailings dams constructed by the downstream or centerline method have engineered dam sections (i.e., the entire section of the downstream dam, and the central and downstream portions of the centreline dam). On the other hand, a dam constructed by the upstream method has only a limited portion along the downstream slope that can be constructed as an engineered dam section (see Fig. 1.9a). The properties of the dam materials within the engineered sections such as gradation, degree of compaction and permeability can be specified, and the material selection, placement and compaction method can be controlled in order to meet the design specifications. In the non-engineered portion of the dam section, the properties of materials are dictated by the manner in which the tailings are discharged and deposited in the tailings impoundment as described below.

Tailings in Non-engineered Dam Section

The typical geometry of a tailings beach surface assumes a concave profile, called "master beach profile" (Blight, 1994). Wang et al. (2011) showed that the overall beach slope steepens with the increasing solids content of tailings discharge (see Fig. 1.7). The properties of deposited tailings on the tailings beach and under the supernatant pond tend to vary both horizontally and vertically according to its location

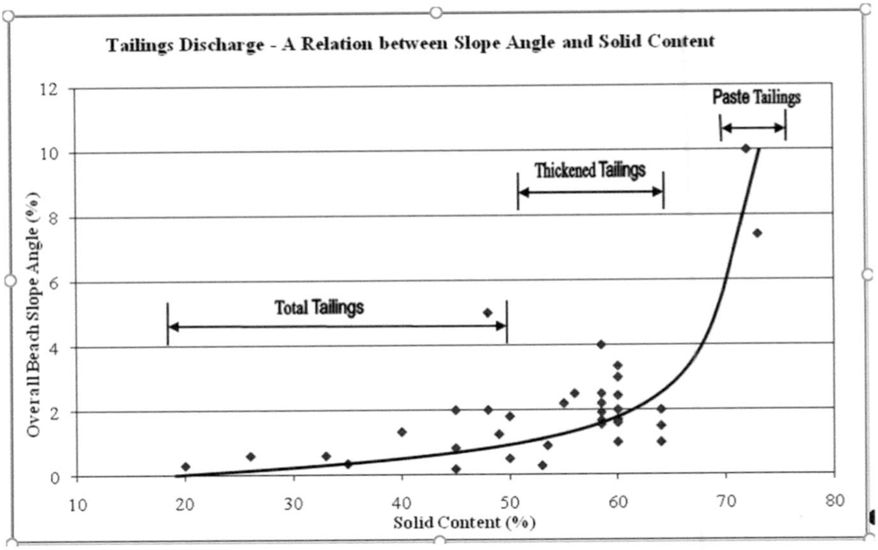

Fig. 1.7 Overall beach slope angle versus tailings discharge solid content (after Wang et al., 2011)

relative to the tailings discharge point and pond level. In general, the deposited material has relatively lower density and higher fines content than the cycloned tailings sand. The tailings material thus deposited, as a rule, is less amenable to design specifications and quality control during construction.

Tailings Sand in Engineered Dam Section

Tailings materials in the engineered dam section are often produced using hydrocyclones, or other thickening processes, and the materials thus produced can be further densified by mechanical compaction to meet the design specifications. Thus, in appropriate circumstances tailings can be improved with reasonable incremental cost to become competent dam construction material. Briefly discussed in the following are cycloned sandfill as dam construction materials.

Cycloned Sandfill

A hydrocyclone is a simple hydraulic device with no moving parts that makes use of the centrifugal principle to separate materials of different sizes (see Fig. 1.8). As whole tailings slurry feeds to a cylindrical chamber, coarse particles spiral downwards through the conic apex at the bottom to form underflow of the coarse fraction. The finer fraction rises to the outlet at the top as overflow. The hydraulic operation is affected by slurry pulp density, pressure and specific gravity. The performance of hydrocyclones can be estimated by their manufacturer, and the estimate can be improved with the testing of actual samples of the whole tailings.

Depending on the grain size distribution of the whole tailings, and the fines content of the desired cycloned sandfill, a whole tailings gradation with 60% fines can be

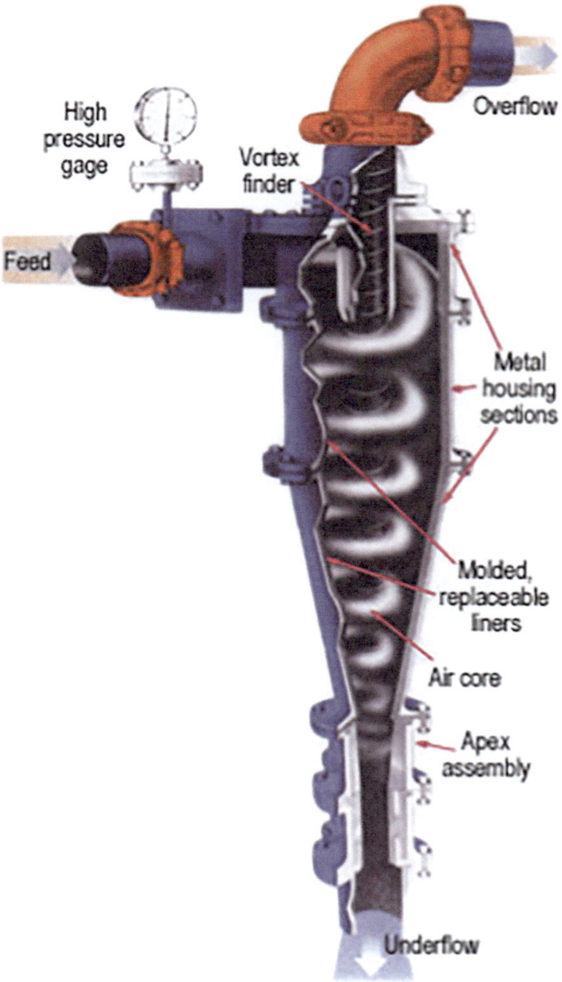

Fig. 1.8 Schematic diagram of a hydrocyclone (hydrocyclone images website)

used to produce about 10–30% cycloned sand fill by weight of a whole tailings. The quantity of cycloned sand production is inversely proportional to the fines content of the whole tailings, and double-stage cycloning, involving primary and secondary cyclones, may be required when clean sand with < 10% fines is to be produced from finer total tailings.

Cyclones can be either fixed or mobile, a fixed cyclone house is used to enclose a cluster of cyclones and often located at one of the dam abutments, from where the cycloned sandfill slurry is pumped either to dam crest or dam slope for placement. Stationary or mobile on-dam cyclones are also used along the dam crest to produce cycloned sandfill either as primary or secondary cyclones. The hydraulic placement of cycloned sand on dam is much more cost effective than the usual excavation-haul-dump-spread mechanical method. Sandfill slurry is end-discharged into hydraulic

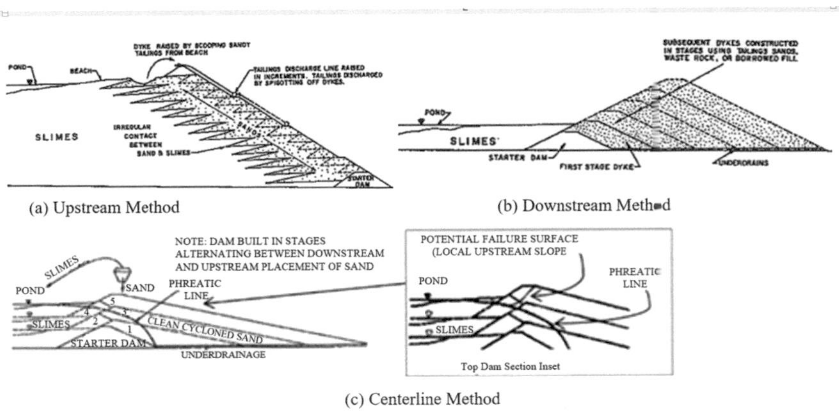

(a) Upstream Method (b) Downstream Method (c) Centerline Method

Fig. 1.9 Schematic sketch of tailings dam construction methods (after Klohn, 1972)

cells and spread by bulldozers. Alternatively, "ropy" sand underflow from on-dam cyclones can be deposited directly on the dam slope to form a newly added layer of damfill. Depending on density requirement of the damfill, systematic mechanical compaction of damfill using bulldozer or compactor could be specified.

1.3 Tailings Storage Facility

There are three main options of tailings storage facilities:

- Underwater disposal in natural water bodies—In the past, tailings has been directly discharged into natural water bodies such as streams, lakes or oceans without worrying about the environmental ramifications of this practice. Currently, underwater disposal is still being practiced under special circumstances. However, its environmental as well as socio-economic impacts tend to be scrutinized more closely with thorough investigation, trade-off consideration and adjudication. Issues involved in this practice include: the loss and replacement of aquatic habitats, the physical, chemical and ecological changes occurring in the receiving waters, and the beneficial effect of prevention of tailings oxidation by its submergence in density-stratified water.
- Underground storage—The primary objective of using tailings as underground mine backfill is to provide a convenient and stable working floor and/or lateral support to underground workings. The backfill also mitigates the long-term ground subsidence problem in the mined-out area. If the backfill is below the long-term water table, prevention of oxidation of acid-generating tailings is another important consideration. Moreover, underground storage would reduce the amount of tailings that would have to be stored above ground.

- Above-ground storage—In terms of on-land tailings storage, an above-ground facility involves storage volume usually at least an order-of-magnitude higher than an underground one. Above-ground storage can be grouped into following three types:
  - dry tailings facility (no peripheral tailings dyke);
  - thickened tailings facility (with low peripheral tailings dyke); and
  - tailings impoundment.

With the advancement of dewatering technology involving high-tonnage, modern thickeners and filters in recent decades, dry and thickened tailings storage have become viable and sometimes advantageous options. These schemes involve stacking and storing unsaturated tailings without the attendant supernatant pond. The peripheral tailings retention dyke could either be eliminated or reduced in height. In general, the cost for these types of facility is much higher (by a factor of 2 or more) than that for the tailings impoundment. They are usually chosen for reasons such as: conservation of water; restriction of storage-site footprint; mitigation of the impact of facility failure; and receipt of favourable public feedback and regulatory permit.

On the other hand, tailings impoundment is generally chosen for the following reasons: large storage volume offered at a site with favourable topography; operational flexibility; and lower cost due to economy of scale. Since major tailings storage facilities involve large tailings dams whose failure having severe impacts, case histories in Chap. 3 are selected from tailings impoundments.

Tailings storage in a decommissioned open pit could be considered as a subset of this option. After the removal of ore body from an open pit, the exhausted pit has already been disturbed by mining operation. Thus, it offers an optional tailings storage space provided that a proper assessment is made on its potential impact on the surface and ground water regime in the surrounding area (EPA, 2000). The open pit formed by coal or oilsands mining is often constrained by the footwall and hanging wall. To close the excavated space for tailings storage, additional dam(s) may be required to span the gap(s) between the two side walls.

1.4 Tailings Dam

Tailings dams of early vintage and those constructed in low seismic and especially dry climate regions are often constructed by a procedure involving a minimum of engineered-fill placement. This was achieved by building a low Starter Dyke and discharging tailings from the top of the dyke until the impoundment was nearly filled with tailings. Then a second small dyke was constructed over the deposited tailings, and the procedure was repeated so that the tailings dam rose gradually in the upstream direction, hence the name "upstream method" was coined (see Fig. 1.9a). Martin and Davis (2000) reviewed some of the basics of upstream tailings dams, including special precautions, as they pointed out almost 2/3 of all tailings dam failures are of dams constructed by the upstream method. This dominant share of failure is a clear warning about the risk involved.

As shown in Fig. 1.9a, an "upstream tailings dam" consists of two main elements: a series of small dykes along the downstream slope of the ultimate dam and bulk tailings sometimes predominantly slimes, deposited upstream of the sloping dykes by hydraulic means such as discharge from large tailings slurry pipes or small-diameter spigot pipes. Thus, the stability of the ultimate dam depends on the nature and competence of the hydraulically placed tailings. Herein lies the main weakness of this type of dam. Only a small portion of dyke-fill constructed as small dykes along the downstream slope is amenable of quality control and quality assurance measures, while the bulk tailings is subject to so many natural and man-made factors that dictate their final engineering properties during their deposition and subsequent settlement and consolidation processes. Thus, upstream tailings dams generally have low seismic resistance, and could even fail under static incremental loading due to spontaneous liquefaction during normal dam raising activity.

The downstream tailings dam construction method is a direct adoption of the conventional water retention embankment dam for the use of retaining tailings (see Fig. 1.9b). This type of dam has always been a safer but more costly type of tailings dam.

The third type of tailings dam is called "centreline dam" as shown in Fig. 1.9c. There are perhaps two perceptions of the centreline dam. On one hand to those who have adopted it, the dam is merely a cost-effective version of the downstream dam. This favourable association becomes intuitive, as outlined by Klohn (1982). Thus, its adoption and obvious relative merits are taken for granted by its adopters. On the other hand, to people or organizations unfamiliar with this type of dam, there exists a perception, which views the method as only a half-measure improvement from the upstream to the downstream dam. This view is understandable, if one takes a cursory look at the respective dam sections corresponding to the three construction methods as shown in Fig. 1.9a–c. To be sure, the dam geometry of the centreline dam does look like somewhere in between those of the upstream and downstream dams. Moreover, the section for the centreline dam appears to be truncated on the upstream side, causing an uneasy feeling to someone unfamiliar with its sound design principles.

The following briefly outlines the cornerstone of the centreline-dam design principles (Lo et al., 2003). By adhering to these principles, the centreline dam is as sound a type of dam as the downstream dam except with much lower cost, when properly designed and constructed.

1.4.1 Upstream Support

By examining Fig. 1.9b, c, it is not difficult to observe that the centreline dam enjoys essentially the same downstream stability as the downstream dam. The upstream stability, on the other hand, requires further elaboration. For a water-retention dam, the upstream water level will rise and fall depending on the operation of the reservoir. A critical phase in evaluating the upstream dam stability is during a rapid drawdown condition. Thus, the upstream slope of a water retaining dam has to be designed for the anticipated range of drawdown levels and drawdown rates, paying due consideration to the drainage and strength characteristics of the upstream embankment zones. However, for a tailings retention dam, the upstream level of stored tailings will only rise and never fall. Moreover, the bulk unit weight of the upstream tailings deposits is not dissimilar to that of the embankment damfill itself. There is little shear stress at the boundary between the tailings and damfill due to essentially balanced weight on both sides. Thus, the upstream support of a centreline dam is provided by the weight of stored tailings deposits. This is analogous to the situation during construction of a concrete diaphragm seepage cutoff wall, where the vertical side walls of excavation are supported by the infilled, thickened bentonite-cement slurry.

Many tailings dams contain a supernatant water pond which rises along with the impoundment level. There often are seasonal fluctuations of the pond level due to the change of various inflows and outflows contributing to the pond water balance. However, in a well designed facility this pond will be separated from the centreline damfill by a tailings beach of sufficient width even in extreme inflow conditions, and poses no significant dam stability concern.

1.4.2 Upper Crest Section

During each raise of the centreline dam crest, there is locally a relatively small unbalanced weight of the embankment dam relative to the upstream tailings deposits in the vicinity of the upper crest section (see top dam section inset in Fig. 1.9c). This unbalanced weight is usually spread over a transition zone, which can be designed to meet the loading conditions specific to a given tailings dam. Design considerations include the relative height difference between the dam crest and the upstream tailings surface, unit weight difference between damfill and tailings, the shearing characteristics of the damfill and adjacent tailings, the width of the tailings beach, the pond water level and site seismicity. Both the design and construction of the upper crest section requires special attention. However, these are important design details inherent to the method, and can be accommodated appropriately. Upon mine closure, the ultimate upstream configuration of the dam and tailings beach can be easily designed to be compatible with the site seismicity and closure scenario involving the permanent spillway, ultimate dam, tailings beach and pond configurations.

Thus, the centreline dam is based on sound engineering and environmental considerations and enjoys inherent economy and safety. Similar to its more familiar downstream-dam counterpart, it should adopt appropriate design criteria according to the specific project requirement. Nevertheless, its design makes more demand on the knowledge, experience and judgement of the designer.

1.4.3 Construction Stages

The tailings dams share many common characteristics as water retention dams, and could be considered as a sub-set of embankment dams. The following three features of a tailings dam, however, allow extra flexibility in its design:

- Unlike a water retention dam, where the level of stored water could go up and down, the level of water stored in the tailings impoundment only goes up, except minor seasonal drop. This feature is favourable for maintaining upstream dam slope stability;
- The stored tailings deposit upstream of the dam has a unit weight comparable to that of the dam fill itself. Thus, the weight of retained tailings offers upstream support for the dam. This support is fully utilized in the design of a "centreline dam" as explained in Sect. 1.4.1; and
- The long construction period of a tailings dam provides an excellent opportunity for the designer to monitor, confirm or modify design assumptions to arrive at a safe yet economical design without relying on over-conservatism. In essence, the time-honoured "observational approach" expounded by Peck (1969) can thus be fully developed especially when dealing with difficult foundation problems.

References

Blight, G. E. (1994). The master profile for hydraulic fill tailings beaches. *Proceedings of the Institution of Civil Engineering, 107*, 27–40.

EPA. (2000). *Workshop on the characterization, modeling, remediation, and monitoring of mining-impacted pit lakes*, Reno.

Jewell, R. J., & Fourie, A. B. (2022). *Paste and thickened tailings—A guide* (3rd ed.). Australian Centre for Geomechanics.

Klohn, E. J. (1982). The development of current tailings dam design and construction methods. *Mineral and Energy Resources, 25*(5), 1–15.

Kossoff, D., et al. (2014). Mine tailings dams: Characteristics, failure, environmental impacts, and remediation. *Applied Geochemistry, 51*, 229–245.

Lo, R. C., McLeod, H. N., & Klohn, E. J. (2003). Centreline tailings dam—Overview of a successful design practice. In *Proceedings of ICOLD Symposium on Major Challenges in Tailings Dams*, Montreal.

Martin, T. E., & Davies, M. P. (2000). Upstream constructed tailings dams—A review of the basics. *Journal of Mining and Environment,* 3–15.

Peck, R. B. (1969). Advantages and limitations of the observational method in applied soil mechanics. *Geotechnique, 19*(1), 171–187.

Wang, B. W., Wu, S., & Plewes, H. D. (2011). Estimate of tailings beach slopes for impoundment design, mining and environ. In *International Conference V*, Sudbury.

Chapter 2
Dam Failures, Emergency Preparedness and Response

Abstract Tailings dam is essentially a subset of embankment dam that adapts to the special characteristics and requirements of the mining industry to provide the storage facility for ever-increasing quantities of fine-grained tailings. Thus, the failure modes of tailings dams are the same as embankment dams. However, because many differences exist in the practice of water-retention embankment dams and tailings dams, their historical performance records do differ. The databases of dam incidents/failures of the respective categories are discussed in Lo (Safeguard embankment dam safety, 2019).

Keywords Dam failure modes · Emergency preparedness · Emergency response · Evacuation order

2.1 Introduction

United States Committee on Large Dams (USCOLD, 1994) conducted a statistical survey of 185 cases of tailings dam failure and incident. The failure case referred to an event involving a dam breach with the ensuing release of impounded tailings, while the incident case referred to an event that did not reach such a serious state. Klohn (1995) made the following observations of the survey:

- Most failures and incidents occurred while tailings impoundments were being in active use as opposed to being inactive (91% of failures, and 82% of incidents). This observation is encouraging for the long-term closure performance of a tailings facility. It suggests that after performing favourably during the active operating phase, the tailings facility would, in general, continue its satisfactory performance after closure, as long as adequate surveillance and maintenance of the facility are rigorously carried out, and the extent of the supernatant pond is reduced significantly;
- The most vulnerable type of tailings dam is constructed by the upstream method (45% of failures, and 41% of incidents), and the next vulnerable type is a tailings dam that also retains water (15% of failures, and 14% of incidents). The former

R. C. Lo, *Tailings Dam Failure*, https://doi.org/10.1007/978-3-030-87738-5_2

suggests that the upstream dam is least robust among all types of dam construction methods, while the latter indicates that the likelihood of tailings incident is relatively higher when water is retained within the impoundment (including post-closure impoundment), especially if water is directly in contact with the dam without a significant tailings beach in between; and

- The most common causes of failures/incidents are slope instability, earthquake and dam overtopping, while issues related to the foundation, seepage and structural components are other important causes. It is worth noting that since tailings dams are earth- and rockfill-dams, once overtopped they would generally fail. Thus, for the overtopping cause, the percentage of failure (16%) is significantly higher than the percentage of incident (4%), while for all other causes, the percentages for failure versus those for incident are comparable.

ICOLD (2001) reviewed 221 cases of tailings dam dangerous occurrences including 185 cases in the USCOLD (1994) database, and had similar observations. Medical causes of the demise of a human are often more than one. Similarly, the demise of a tailings dam also usually has multiple causes, including primary and secondary ones. Moreover, these causes could interact, hastening the process of taking down the vulnerable dam. In Sect. 2.2, these causes are explored one by one.

Section 2.3 describes the impacts of tailings dam failure, and Sect. 2.4 examines both natural and human factors that affect dam safety.

Besides running a normal and cost-effective dam operation, the mine operator needs to keep vigilance in anticipating which component of the tailings dam system could go wrong. What are the potential consequences of such a mishap? How to respond to the mishap and minimize its consequences? Thus, it is critical to have a well-thought-out emergency response plan, detailing concrete actions to be taken by personnel at all levels as well as to have well-defined internal and external emergency communication protocols. Section 2.5 describes key considerations in dealing with emergency situation for safeguarding dam safety and general public whose lives and livelihood are threatened in a potential dam breach.

An emergency can influence the operation of the entire mining organization, which is also the responsible entity for all consequences associated with the emergency. To resolve the emergency expediently with the optimal outcome, the mine operator needs the cooperation of governmental regulators, emergency responders, technical consultants, construction contractors, and the public. Thus, the mine should prepare response procedures and communication protocols in normal times so that it could take appropriate actions readily in case of an emergency.

The response plan should cover a comprehensive overview of all potential scenarios as well as appropriate actions to manage and control dam safety issues in the specific type of incident. The most critical action to safeguard public safety is to review the possibility of a dam breach and to allow adequate time for the government emergency responders to issue and execute the evacuation order. Finally, training, regular practice, and documentation update are essential for keeping the response plan a living instrument to ensure dam safety.

2.2 Dam Failure Modes

The analysis of potential failure modes of a dam is a very powerful tool used in dam design and safety assessment. It is also used in a forensic investigation of dam incidents and failures.

2.2.1 *Slope Instability*

The behaviour of following soils and clay shale in response to static load needs to be considered carefully to guard against unsatisfactory or even dangerous dam performance:

Loose Silt and Sand

Loose cohesionless silt or sand tends to contract when it is sheared. If the shear load is applied so rapidly that the pore water cannot drain out, the pore water will be subject to a pressure increase. Correspondingly, the effective stress existing at soil grain contacts will be reduced by the same amount. In turn, the soil shear-resistant strength derived from friction among soil grains will also be reduced. This loss of shear strength could lead to a flow slide of the affected dam, and the phenomenon is called static liquefaction. The recent rash of large upstream dam failures and their forensic investigations provide detailed evidence regarding the extreme rapid progression of the final dam collapse as discussed in Chap. 3.

Sensitive Clay

The undrained shear strength of sensitive clay in a remolded state is only a fraction of that in an undisturbed state. The ratio between the strength of the undisturbed state to that of the remolded state is called the sensitivity ratio of the clay. The soil acquires its sensitivity from the history of its physico-chemical development from initial deposition until the present.

Highly Plastic Clay and Shale

The plasticity of highly plastic clay and shale is related to their content of platy montmorillonite clay particles. Their available shear strength decreases with the increase of shear deformation along a potential failure plane due to the reorientation of platy clay particles along the plane to a much lower residual value than the peak value corresponding to a relatively small deformation. This could lead to slope failure or rapid large deformation, if the safety factor of the deformed mass drops down to a value approaching unity.

2.2.2 Dam Overtopping

Tailings dams, being a subset of embankment dams, offer little resistance to water erosion, when the tailings impoundment water level rises above the dam crest. Once overtopped, a tailings dam is often subjected to the down-cutting action of the continual erosion force of the released mixture of water and tailings. Unless the dam's down-cutting process is arrested timely by emergency repair and/or safe water withdrawal from the impoundment, the dam would be doomed.

2.2.3 Earthquake

Earthquake loading has dual effects on the deformation and stability of a dam. During an earthquake, the dam is subject to horizontal and vertical inertial forces in addition to the ever-present gravitational force. Moreover, the pore water pressure will increase due to earthquake shaking. As a result, the dam deformation will increase and dam stability will reduce. If the dam fill or foundation material is susceptible to liquefaction, a flow slide could be triggered.

2.2.4 Foundation Instability

If the foundation is involved in a dam instability or failure, the case is classified as a foundation failure. Thus, foundation assessment is part and parcel of the task of dam-site selection. The irony of an upstream dam is the fact that the bulk of such a dam is founded on non-engineered material produced by tailings operation subject to the vicissitude of tailings discharge, flow and deposition (see Fig. 1 9a). A favourable natural dam site could be spoiled by the deposition of undesirable tailings slimes scattered in an unknown fashion within the foundation footprint of the ultimate dam as revealed in many case histories in Chap. 3.

2.2.5 Seepage/Internal Erosion

Internal erosion could cause unexpected and uncontrolled seepage loss or even dam breach through the "piping" mechanism (Fell et al., 2003). Defence measures consist of proper internal zoning of the tailings dam section as well as maintaining gradation compatibility at the boundaries between various dam zones and the dam/foundation contact (Foster & Fell, 2001). Internal erosion along easy passageways such as cracks or the peripheral zone of buried pipelines should be specially guarded against (Cedergren, 1977; Hirschfeld & Poulos, 1987).

Fell and Fry (2007) edited a book on the Internal Erosion of Dams and Their Foundations based on a workshop held in Aussois, France. It presents an overview by the two editors on the State of the Art of the physical processes of internal erosion and the framework for the safety assessment of internal-erosion incidents related to water retaining structures. Moreover, papers on laboratory testing, case histories and practical experiences of international dam agencies on the subject are included.

Generally, the demand for the quality of filter materials is less stringent for a tailings dam than for a water dam. The availability of a wide tailings beach during operation significantly reduces the prevailing seepage gradient within the dam. The eventual elimination or reduction of the supernatant pond after closure is another favorable factor. However, these favourable factors have to be clearly stated as design criteria in the design-base memorandum. Furthermore, these criteria should be confirmed at all times during tailings operation and facility decommissioning to prevent an unexpected dam incident or failure due to the discrepancies between the real condition of the as-built tailings dam and the fictitiously favourable design criteria.

2.2.6 Failure of Components of Flood-Control Structures

In a closed water management system, flood water is stored within tailings impoundment. Provision of sufficient flood-storage capacity for the design flood at all times is critically important to guard against dam overtopping. In an open system, hydraulic structures, such as sloped or vertical decant intake and open or closed spillway, are used to safely convey flood water from the impoundment to the downstream watercourse away from the tailings dam. The structural integrity and smooth operation of all components of the flood-control structures with adequate water-carrying capacity are to be maintained vigilantly during flood seasons. Regular inspections and testing of these structures are necessary to ensure their proper function in flood seasons.

2.3 Impacts of Dam Incident/Failure

Impacts of tailings incident or failure on the affected communities include:

- loss of life and injuries;
- loss of property and infrastructure; and
- loss of livelihood, which could be caused by loss of employment due to mine shutdown and/or flood damage and environmental pollution.

Depending on the nature of the tailings mishaps, the impact on the environment could progress either slowly or rapidly. If the problem is geo-chemical with no dam breach, the impact tends to progress slowly. Remedial measures could be implemented with ongoing monitoring to confirm their effectiveness or for further improvement. However, for a sudden failure of the retention dam with the ensuing release of tailings and/or water, the impact develops rapidly. It requires the fast deployment of personnel and equipment to engage in emergency response measures, as described in Sect. 2.5, to minimize downstream community and environmental damage, and to improve the effectiveness of restoration work.

It is important to point out the precarious state of those victims who would suffer the fate of losing their livelihood or even life due to a dam-breach failure, yet often would be powerless to protect themselves in the current prevailing socio-economic conditions. In the case histories presented in Chap. 3, their various experiences illustrate vividly the compelling need for us to turn the tide of the serial dam breaches that occurred in recent years.

2.4 Factors Affecting Dam Safety

In the real world, Murphy's Law dictates that sooner or later there would be some mishaps occurring somewhere. Two main potential causes that contribute to dam mishaps are natural and human hazards. This section examines these hazards so that we will be better prepared to minimize their influence on dam safety.

2.4.1 *Natural Hazards*

Natural hazards most pertinent to the safety of an embankment dam are floods (including those resulting from climate change) and earthquakes. These hydrological and seismic hazards must be carefully evaluated and designed for. Appropriate design criteria must be established which are compatible with the project site setting and the severity of consequences of a dam failure to the downstream communities and environment. Other hazards, such as fire (for coal waste), landslide, debris flow and avalanche, may also need to be assessed when appropriate.

2.4.2 *Human Hazards and Factors*

A well-organized, efficient dam-safety management system should address the following human hazards and factors that could potentially undermine any weaker links of the system:

- Individual and/or institutional blind spots and culture biases—knowledge, experience, vigilance and critical review by independent experts might counter these deep-rooted issues.

- Internal and/or external constraints that may result in the gradual reduction of margin of safety, loss of safety control and sudden dam failure—common constraints include insufficient resources in time, funds and expertise.
- Break-down of management system—changes in the management system are inevitable over the long life of a tailings dam, including changes in ownership, organization and personnel—proper documentation, the overlap of key personnel during the transition period and effective audits might minimize this threat.
- Errors and omissions—although "to err is human", built-in defence mechanisms are needed to catch errors in time before serious incidents occur. Quality management systems, peer review, risk assessment and management, and study of failure case histories are some of these defence mechanisms. Errors and omissions could creep in during any stage including design, operation/construction and closure. They could also occur in any organization, including designer, consultant, operator or regulator.
- Sabotage and terrorist activity—this includes all deliberate activities to damage or destroy dam facilities. Proactive security surveillance and vigilant control of site access and activities commensurate with prevalent site conditions are needed to guard against this threat from people with mal-intent.

2.5 Emergency Preparedness and Response

The dam operator bears the responsibility of taking immediate remedial actions to deal with a dam incident as well as to report timely to the regulatory agencies and emergency responders to alert people who could potentially be impacted. Consulting firms and contractors are often promptly engaged to quickly diagnose the correct cause of the problem and to identify appropriate response measures. These measures include stabilizing the problem, if practicable, and/or initiating downstream evacuation procedures, often through a governmental emergency agency. In the case of an abandoned tailings facility, the government environmental or emergency agency often becomes the focus of forming a multi-agency response centre. The response centre coordinates appropriate actions that are necessary for dealing with the emergency, repairing the damages and/or restoring the existing facilities.

2.5.1 Response Procedures and Communication Protocol

The response procedures are to be documented with an action plan for addressing: the planning and implementation of responsibilities and decision-making framework; equipment, materials and personnel required and available for emergency response; emergency scenarios; general response to the identified events; and communication plan covering internal, regulatory, community, public and media.

2.5.2 Potential Outcomes

Not all dam incidents lead eventually to a dam breach or escape of large amounts of stored water (and tailings). However, most embankment dams are vulnerable to erosion by the flow of water (and tailings) over the dam crest. Thus, emergency dam repairs need to focus on preventing/managing overtopping, including water removal from the impoundment to restore and increase the freeboard (the elevation difference between the dam crest and pond level).

Internal (piping) erosion of a dam may also lead to a dam breach, if the eroded internal conduit(s) reaches the reservoir or supernatant pond and enlarges with time. In some favourable situations where a less erodible rockfill zone of a tailings dam is present, the erosion process could potentially be stopped by itself, as the hydraulic gradient of seepage through tailings reduces with the escape of some pond water at shallow depth. Or, the erosion process can be stopped by an emergency repair.

In the case of slope failure involving damfill and/or foundation, loss of freeboard and escape of water through the failed section could eventually lead to a dam breach, unless the freeboard is restored promptly. Only when the dam breach is prevented, can one then plan for a permanent repair and restoration of the existing dam.

2.5.3 Evacuation Order

Concurrent with taking immediate remedial measures to prevent a dam breach, the dam operators should not overlook their critical responsibility of assessing the situation realistically and conservatively, and allow sufficient time for an orderly evacuation of the downstream communities to escape potential drowning by water or burying by tailings mud. Close liaison should be maintained with the government emergency agency so that the authority would have adequate time to initiate and execute the evacuation procedures to avoid loss of life. While unwarranted false alarms are to be avoided, appropriate conservatism is needed to make this all-important decision call regarding timely evacuation. Figure 2.1 illustrates a step-by-step emergency action plan to be taken by the dam operator. Regular dry runs of this plan should be practiced both internally and externally to improve it by rooting out any weaknesses. Case histories in Chap. 3 depict some unfortunate situations, where the preciously short window of opportunity for safe evacuation was either not available due to late detection of dam failure in progress or squandered by key personnel who failed miserably in their responsibility to report timely to the government regulators and emergency responders.

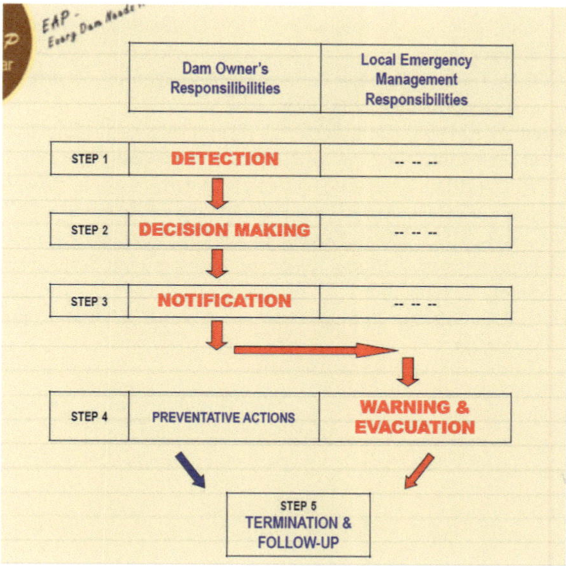

Fig. 2.1 Flow chart illustrating execution of emergency action plan (from NRCSI-ASDSO, 2011)

2.5.4 *Training, Practice and Documentation*

The emergency-responder information package is to be updated and distributed at least annually. The information package includes:

- emergency response notification listing;
- mine site boundaries and fire coverage;
- site facilities and access roads;
- hazardous material inventory; and
- locations of emergency-response equipment and materials, etc.

Training and exercise are to be conducted at least annually. In addition to tabletop exercises, real enactment of emergency scenarios with the participation of all relevant responders and affected residents should be held as well.

References

Cedergren, H. R. (1977). *Seepage*. Wiley.

Fell, R., & Fry, J.-J. (2007). *Internal erosion of dams and their foundations*. Taylor & Francis.

Fell, R., Wan, C. F., Cyganiewicz, J., & Foster, M. (2003). Time for development of internal erosion and piping in embankment dams. *Journal of Geotechnical and Geoenvironmental Engineering, 129*(4), 307–314.

Foster, M., & Fell, R. (2001). Assessing embankment dam filters that do not satisfy design criteria. *Journal of Geotechnical and Geoenvironmental Engineering, 127*(5), 398–407.

Hirschfeld, R. C., & Poulos, S. J. (1987). *Embankment-dam engineering* (Casagrande Volume). Krieger Pub.

ICOLD. (2001). Tailings dams risk of dangerous occurrences—Lessons learnt from practical experiences. *ICOLD Bulletin, 121.*

Klohn, E. J. (1995). Tailings dams—Their evolution to major hydraulic structures and the importance of their safety. In *Keynote Address to Canadian Dam Safety Conference*, Banff, Alberta.

Lo, R. C. (2019). Safeguard embankment dam safety. In *Proceedings of Symposium on Sustainable and Safe Dams Around the World*, Paper 65, Ottawa.

NRCSI-ASDSO. (2011, November 15–16). *Emergency action planning—Training seminar*, Portland, Oregon, National Resources Conservation Services Institute—Association of State Dam Safety Organization.

USCOLD. (1994). *Tailings dam incidents*. USCOLD Committee on Tailings Dams, Denver, Colorado.

Chapter 3
Representative Case Histories and Lessons Learned

Abstract Case histories of dam incidents and failures contain invaluable sources of practical and insightful information to guide practitioners conducting normal operations, assist regulators in establishing safety-conscious practices and enforcing rules and regulations to safeguard people's life and property and the environment free from pollution. The selection of representative case histories presented herein is based on databases available to the author, including: (1) collections by dam-building institutions such as International Commissions of Large Dams (ICOLD) and its national commissions; (2) technical papers; (3) forensic reports tasked by government agencies and dam owners; and (4) news articles. These cases include common dam failure modes: static slope instability; static liquefaction; dam overtopping; seismic slope instability; foundation instability and seepage-induced (piping) internal erosion.

Keywords Case histories · Lessons learned · Dam incidents · Dam failures

3.1 Introduction

The select cases occurred in developed as well as developing countries scattered around the world during different stages in the life span of respective dams. In the following section, select cases are presented in chronological order. Some of these cases played an important role in influencing the evolution of tailings dam practice. These influences are highlighted in relevant cases.

Twenty-one cases are presented in the following section. All but one involved some incidents or failures due to a variety of reasons. The exceptional case, Case 21, where the possibility of foundation instability was recognized early, and serious efforts have been taken to manage this potential risk to date. A summary of all these cases is given at the end of cases presentation to facilitate easy referencing.

R. C. Lo, *Tailings Dam Failure*, https://doi.org/10.1007/978-3-030-87738-5_3

3.2 Case 1 March 28, 1965 El Cobre (El Soldado) Tailings Dams Seismic Liquefaction Flow Failure, Chile

The failures of the El Cobre (El Soldado) Mine's tailings impoundments together with other impoundments in Chile during the March 28, 1965 La Ligua earthquake served as an impetus to focus the mining industry's attention on the vulnerability, under seismic loading, of tailings dams constructed by the upstream method. Cohen and Moenne (1992) reviewed the tailings practice at El Cobre (El Soldado) mine both before and after the 1965 earthquake, while Dobry and Alvarez (1967) provided an overview of the La Ligua earthquake impact on Chilean tailings dams including El Cobre tailings dams. The March 28, 1965 La Ligua earthquake was an interplate subduction event of magnitude 7 to $7^{1/4}$ with a focal depth of 61 km. The epicentral distance to the El Cobre (El Soldado) mine was about 40 km, and the earthquake intensity there was VIII to IX on the Mercalli scale. Figure 3.1a, b show, respectively, the back and front view of the failed Old Dam, while Fig. 3.1c shows the cross-section of the dam in a distorted scale (vertical scale = 4 × horizontal scale). At the time of the 1965 earthquake, El Cobre (El Soldado) had three tailings impoundments: Viejo (Antigua), Chico and Nuevo. Some of the relevant information on these impoundments and their earthquake performance is summarized in Table 3.1.

All three tailings dams were constructed by the upstream method. At the time of the earthquake, both the Viejo and Nuevo impoundments were fully operational for tailings storage, while the Chico impoundment was only used intermittently when the other two dams required maintenance. The intermittent use of the Chico impoundment allowed time for tailings to drain and consolidate, resulting in a lower phreatic surface within the dam. The more favourable seismic performance of tailings impoundments that are either not in use or only in partial use during the earthquake such as the Chico tailings dam has been the general experience elsewhere as well. The suddenly released 2.4 million m^3 of tailings from the two failed impoundments rushed downstream along the valley, travelling 12 km in a few minutes. The tailings flow slide wiped out part of El Cobre town and killed more than 200 people.

Because of the widespread tailings impoundment failures caused by the La Ligua earthquake, the Chilean Ministry of Mining tightened its regulations on tailings dams in 1970 by the promulgation of Supreme Decree No. 86. The lessons learned from this earthquake as incorporated in the 1970 Chilean regulations had a profound influence on the advancement of tailings engineering not only in Chile but also around the world, especially in seismically active regions.

At the El Soldado mine, two tailings impoundments were constructed during the transition period of 1965–1970: No. 2 and No. 3 impoundments further up the valley from the Chico impoundment. Although these impoundments were still constructed using the upstream method, on-dam cyclones were used to obtain coarse sands for dam building. In 1970, with Supreme Decree No. 86 in effect, El Soldado started the construction of the modern No. 4 tailings impoundment nearby using the downstream method. A cyclone station was located at the left abutment, and the cycloned sand underflow was transported to the dam using positive displacement pumps for

(a) Back View

(b) Front View

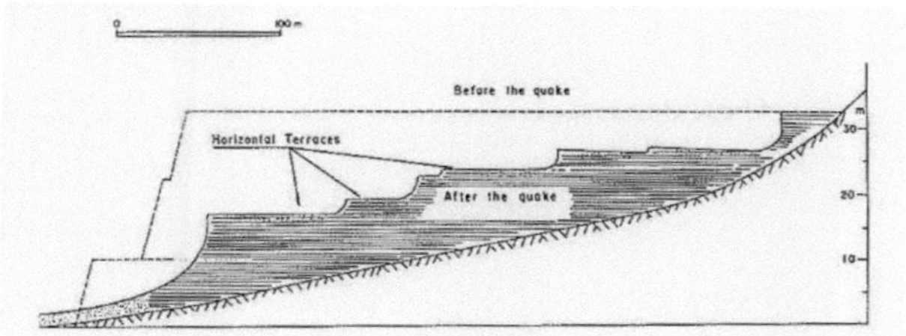

(c) Pre- and Post-Failure Cross Sections (Scale: Vertical = 4 x Horizontal)

Fig. 3.1 Configurations of failed Old Dam (from Dobry & Alvarez, 1967). **a** Back view. **b** Front view. **c** Pre- and post-failure cross sections (scale: vertical = 4 × horizontal)

Table 3.1 El Cobre (El Soldado) tailings dams performance in 1965 La Ligua earthquake (after Cohen & Moenne, 1992)

Name of tailings impoundment	Year when operation began	Maximum dam height (m)	Tailings storage (m^3)	Tailings lost in flow slide (m^3)	Earthquake damage
Viejo (Antigua)	1930	32–35	4.25 million	1.9 million	Dam breach due to liquefaction
Chico	1930	26	0.985 million	–	Downstream slope cracked and slumped
Nuevo	Dec 1963	19	0.5 million	0.5 million	Dam breach due to liquefaction

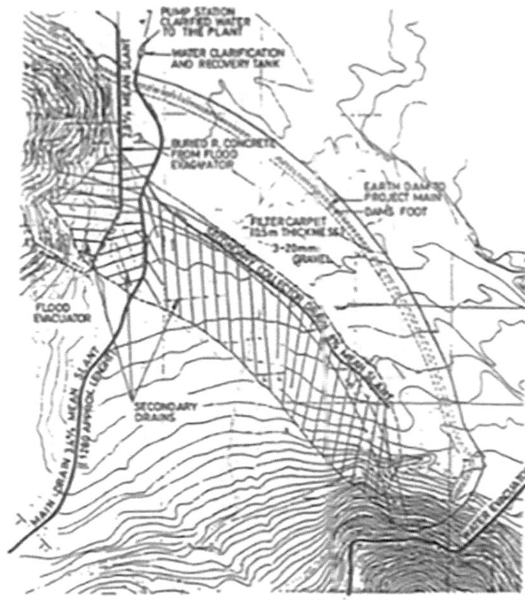

Fig. 3.2 Plan view of No. 4 tailings dam (from Cohen & Moenne, 1992)

the dam building. A decant line was used to reclaim water from the supernatant pond to beyond the downstream dam toe for pumping back to the mill. The underdrainage system consisted of finger drains under the upstream shell and ultimate crest and a 50–60 cm thick blanket drain under the downstream shell. For surface water management, there was a diversion channel on the right abutment, and a buried reinforced concrete spillway on the left abutment (see Fig. 3.2). Dam instrumentation includes 4 Casagrande-type standpipe and 29 electric piezometers and 3 accelerographs (one each at the crest, downstream toe and left rock abutment).

Since 1965, the El Cobre (El Soldado) mine experienced two other large earthquakes, the seismic performance of the tailings impoundments is summarized in Table 3.2.

During the 1985 earthquake, the relatively favourable performance of the old tailings impoundments may be attributable to the fact that they were not actively in use.

At No. 4 tailings impoundment, the damages were limited to the upper portion of the downstream slope under active hydraulic-fill construction and the crest area due to amplified ground motion and the relatively loose and saturated upstream tailings beach. A recent satellite imagery of El Soldado Mine is shown on Fig. 3.3.

Table 3.2 El Cobre (El Soldado) tailings dams performance in 1971 and 1985 earthquakes (after Cohen & Moenne, 1992)

Earthquake	Older tailings impoundments	No. 4 tailings impoundment
M = 7.4 Eq. at La Ligua on July 9, 1971	Survived the earthquake, but no other information available	Max. dam height = 15 m, cracks developed on the downstream slope in areas between finger drains, with a relatively high piezometric level
M = 7.5 Eq. near San Antonio on March 3, 1985	Cracks developed on the crest, downstream slope and upstream tailings beach of No. 2 tailings impoundment	Max. dam height = 45 m, sand volcanoes and depressions developed on tailings beach, cracks on raised earthfill (borrowed earthfill to cope with cycloned sand shortage) dam crest and shallow slope slumps in the upper portion of cycloned sand downstream slope under hydraulic-fill construction

Fig. 3.3 Recent satellite imagery of El Soldado mine (from Google Earth Pro 2020)

Lessons Learned

The performance of El Cobre (El Soldado) tailings dams in three large earthquakes: 1965 magnitude 7 to $7^{1/4}$ and 1971 magnitude 7.4 La Ligua events and 1985 magnitude 7.5 San Antonio event has the following two important lessons for Chile and other countries as well:

1. Seismic behaviours of tailings dams vary depending on the design and construction method and operating condition of the specific dam. A downstream dam is far more robust than an upstream dam. The performance of an upstream dam in an earthquake depends on the water table within the dam. The lower the water table, the safer the dam.

2. Chile tightened its regulations on tailings dam practice with the promulgation of Supreme Decree No. 86 in 1970, including the banning of upstream dams. For a highly seismic country, the Chilean government adopted the policy to avoid upstream tailings dam back in 1970 is both wise and fore-sighted. This lesson seems to be very difficult to take to heart, as other case histories in this chapter will reveal.

As more recent case histories in this chapter will show that the upstream method may prove to be unsafe even without additional seismic loading. The phenomenon of static liquefaction can collapse a tailings dam constructed by the upstream method. The phenomenon can occur without any warning and within a few minutes it can transform a solid looking "dam" into a turbulent debris flow as recorded in a video in Case 18.

3.3 Case 2 September 12, 1969 Twin Creek Tailings Dam—Piping Internal Erosion Incident, Canada

Twin Creek Tailings Dam is a rockfill dam located near Greenwood, British Columbia. The dam was brought into operation in 1967. The design called for the placement of a filter zone consisting of plastic ablation till materials, a mixture of sand, gravel and cobbles with a plastic silt binder, on the upstream face of the dam. The dam was mainly constructed of waste rockfill with rocks up to 450 mm in size. In addition, cycloned sandfill was placed over the filter zone, and a spigotted tailings beach was developed further upstream. The design relied on maintaining, at all times a generous beach width between the supernatant pond and the cycloned-sand face of the dam.

The dam suffered a sudden "piping" failure on September 12, 1969, which caused the loss of large volumes of water and tailings (see Fig. 3.4). Early in the morning, about 3:00 a.m. of that day, a "pipe" developed at a point where pond water came in direct contact with the cycloned-sand face of the dam. The dam watchman heard trickling water as he was standing on the crest of the rockfill section. By 7:00 a.m., the main flow of water occurred at an estimated peak flow rate of 180 m^3/min. For the emergency repair, the mine mobilized mine shovels and trucks to plug the leakage area with filter materials. The flow rate was reduced to about 3 m^3/min by 11:00 a.m. In total, about 38,000 m^3 of water was released resulting considerable property damage downstream. There was no fatality in this incident.

Subsequent investigations indicated that:

- The filter materials for the second and third lifts of rockfill were end-dumped from the top of the lift rather than placed in thin lifts, leading to segregation of filter materials;
- In the failure area, the tailings beach was poorly developed with free water in the pond getting close to the upstream face of the dam. Due to the relatively high seepage gradient in the area, sand was carried by seepage flow into the rockfill,

Fig. 3.4 Internal erosion "piping" of Twin-Creek Tailings Dam (from Klohn, 1979)

resulting the formation of "piping" within the sand zone. In turn, the "pipes" thus formed created conduits from the supernatant pond to the large voids of rockfill section of the dam; and

- Subsequent review of the dam construction record revealed that sinkholes and fluctuations of seepage flows occurred twice in the summer in this area prior to the major "piping" failure. The operators were not alarmed sufficiently by these occurrences to rectify the dam deficiencies, although they did infill the sinkholes with tailings.

The remedial solution was to push the supernatant pond further away from the dam face by 300 m. This was accomplished by building a second dyke across the tailings beach about 100 m upstream of the rockfill section. A spigotted tailings beach was constructed upstream of the newly added dyke. The space between the dyke and main dam was infilled with cycloned sandfill. The remedy worked well and the operation of the tailings facility continued until its eventual decommissioning in 1979.

Lessons Learned

The mine made following mistakes in its dam raising operation:

1. It failed to recognize the importance of maintaining a tailings beach with sufficient width between the supernatant pond and the earth and rockfill tailings dam to reduce seepage gradient;
2. It caused segregation of filter materials in local areas due to poor construction procedure; and
3. It did not recognize the significance of sinkhole formation and seepage fluctuations occurred earlier prior to the major piping erosion incident.

The dam was repaired by lengthening the tailings beach width substantially, and decommissioned subsequently 10 years later.

3.4 Case 3 February 26, 1972 Buffalo Creek Coal Mine Tailings Impoundments Overtopping Failure, USA

Buffalo Creek Valley is a narrow valley with a width ranging from 400 to 600 ft. and steeply sloping valley walls. It is situated at the easternmost end of the Logan County in West Virginia, roughly 75 miles southwest of Charleston. The stream flows in a west by southwesterly direction for 17 miles before reaching its confluence with the Guyandotte River at the town of Man. A series of three coal waste banks (embankments) serving the need of a coal preparation (washing) plant were the focus of this dam failure incident. Figures 3.5 and 3.6 show the plan view of these waste banks located along the Middle Fork tributary of the Buffalo Creek. Davies (1973) gave an overview of this case history, while Wahler (1973) provided a detailed report on the technical investigation of the dam failure. The official report from the Governor's Ad Hoc Commission of Inquiry is given in West Virginia (1973).

The burning Coal Waste Bank in Fig. 3.5 was constructed first, followed by the construction of Dam No. 2 and No. 3 further upstream around 1960s. The state of regulations, design and construction practices for these coal waste banks did not meet modern standards, and possibly did not meet the standards of the day for dams, probably because these facilities were not originally used for storing water. The waste banks did not have proper hydrological and geotechnical design, and embankment fills were constructed loosely over unprepared foundation (subsequent explorations found that sludge deposits beneath Dam No. 3 ranged between 40 and 100 ft. in thickness).

Starting from 1960, the mine was not allowed to discharge coal wash water from preparation plants directly to natural streams. The mine chose to use the coal waste banks to retain coal-wash water in order to clarify the seepage water for stream discharge. Later as strip-mine coal was processed through the coal preparation plant, the waste banks contained more clayey fines, and became more water tight. Thus, an unfortunate pre-condition was set for potential future failure, by transforming the poorly constructed, non-water storing coal waste banks to dams for retaining large ponds.

Figure 3.7a shows the tailings impoundment upstream of Dam No. 3. The dumping of coal waste for constructing the dam was carried out across the valley from the right abutment on the mine access road toward the left abutment. The dumping was done in the form of a single lift which ranged up to 60 ft. thick. Once the dam had been completed across the valley, dumping was mainly across the back side of the dam which steadily grew upstream. Coal waste was end dumped from trucks in closely spaced piles from 4 to 7 ft. high, and then graded into layers from 2 to 4 ft. thick. Trees in the path of the dam construction were not removed but covered by dumped waste. The pool area also was not cleared of vegetation. The sludge on which the waste was dumped over was only partially displaced and much of it formed the foundation of the dam. The front face of the dam sloped 37°, and the back sloped 32°. About 800,000 tonnes of coal waste had been dumped within the dam proper to construct the dam. Altogether, Dam No. 3 contained about 10 million cubic feet of coal waste

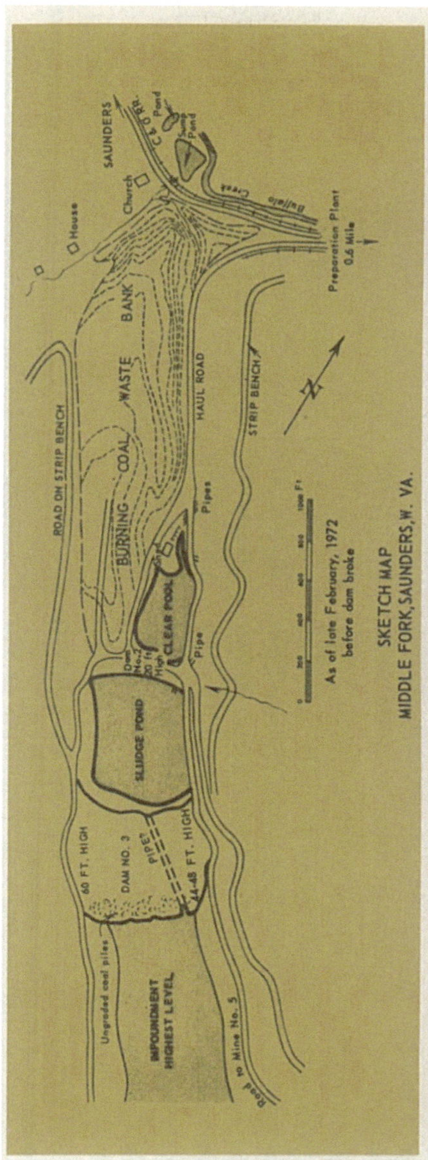

Fig. 3.5 Middle Fork coal waste impoundments (from Davies, 1973)

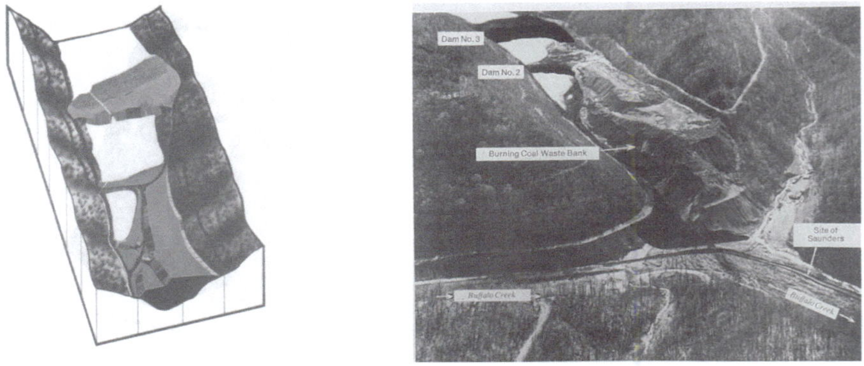

(a) Schematic Sketch of Impoundments (from Wahler 1973)

(b) Bird's Eye View of Impoundments (from Davies 1973)

Fig. 3.6 Coal waste impoundments prior to dam 3 failure. **a** Schematic sketch of impoundments (from Wahler, 1973). **b** Bird's eye view of impoundments (from Davies, 1973)

above the sludge level of Pool No. 2, and another 7 million cubic feet of coal waste displacing the sludge below that level.

During the construction of Dam No. 3, a major portion of the refuse sank in February, 1969, and displaced some of the sludge upon which it was being built. Again, in February or March, 1971, the downstream side of the left abutment of Dam No. 3 slumped. The size of the slump was estimated to be 150–200 ft. wide across the face of the dam and 20–30 ft. deep from the face back. It was observed that black water was boiling into Pool No. 2 from the downstream side of the dam. The pool level in the reservoir behind Dam No. 3 was 20 ft. below the crest at that time. There was an "emergency spillway" installed in 1971 at Dam No. 3. The "spillway"

(a) Impoundment Upstream of Dam No. 3

(b) Aerial View of Flood Debris

Fig. 3.7 Tailings impoundment and flood debris (from Davies, 1973). **a** Impoundment upstream of Dam No. 3. **b** Aerial view of flood debris

was a length of butt-welded steel pipe, 24 in. in diameter and ¼ in. in thickness, placed diagonally across the right side of Dam No. 3, with no collars or baffles, at a depth of about 7–8 ft. below the graded crest.

During the 72-h period immediately preceding the dam breach flood, precipitation averaged 3.7 in. at Logan (corresponding to a 2-year-return event), and about half of the rainfall (1.9 in.) at Logan fell in the 24 h that preceded the flood. At about 8 a.m. February 26, 1972 sudden release of 17.6 million cubic feet of water and suspended coal waste occurred through a breached Dam No. 3. The instantaneous release of the water dropped 253 ft. in elevation to reach the Buffalo Creek valley floor. The flood wave washed out two additional waste banks and impoundments in its path, and tore off the corner of a burning mine-refuse bank before cascading directly into the western slope of Buffalo Creek Valley, one-half mile away. The 10–20 ft. high flood wave traveled the valley at an average velocity of more than 7 ft. per second reaching the settlement of Man in 3 h.

During the flood, 125 people lost their lives, and a total of 4000 people were left homeless. Five hundred and seven houses were lost or demolished; 44 mobile homes were destroyed. Another 273 houses were severely damaged; while nearly 663 more houses suffered damage to varying degrees. In addition, 30 business establishments, 1000 automobiles and trucks, 10 bridges, and power, water and telephone lines were all destroyed. The county road and rail lines servicing the valley's coal mines were severely damaged. The photograph in Fig. 3.7b shows typical damage wrought by the flood. Property damage was estimated in excess of $50 million, while highway damage exceeded $15 million. As the result of this tragedy, US Congress enacted the Surface Mining Control and Reclamation Act (SMCRA, 1977), which has been the primary federal law that regulates the environmental effects of coal mining in the United States.

Prior to the dam breach there were about 12 h or even longer time available for the downstream residents to evacuate, if only the signs of imminent dam failure were recognized by the responsible company so that the public could be pre-warned. Unfortunately, this precious time was lost, while the company personnel were busily engaging in fruitless activities to divert flood water from the impoundment.

In a Governor's Ad Hoc Commission of Inquiry, Deputy Sherriff, Mr. Mutters, stated, that "If (the company) had known that dam was going to break, and informed us of that, then, hell, I don't think anybody would have got drowned. Chances are we could have got everybody out, me and the other patrol car" (West Virginia, 1973).

Lessons Learned

Buffalo Creek Coal Mine Tailings Impoundments Overtopping Failure is an important case history for the following two main reasons:

1. The cascade failure is a significant tragedy in terms of loss of lives and properties as well as infrastructures and environmental damages. After three days of heavy rainfall, a cascade failure of Dam No. 3, Dam No. 2 and Coal Waste Embankment occurred, as the result of the overtopping of Dam No. 3. Moreover, the original function of these failed structures was not for storing water. However, long before

these structures failed, they had been used to store water due to historical development of coal mining practice. Finally, regretfully the mining company failed to report the impending structures collapse to authorities to allow the execution of an emergency evacuation of downstream population at risk.

2. From the regulatory' perspective, the Buffalo Creek Coal Mine Impoundments collapse provided a good opportunity for the US Congress to plug a critical regulation gap in the promulgation of Surface Mining Control and Reclamation Act (SMCRA, 1977). It established the Office of Surface Mining Reclamation and Enforcement (OSMRE) to coordinate the regulation efforts by the Secretary of the Interior and the States.

3.5 Case 4 1978 Hozukizawa Tailings Impoundment Seismic Liquefaction Flow Failure, Japan

Mochikoshi Gold Mine's beneficiating (concentration) plant and Hozukizawa tailings impoundment are located in Izu peninsula, Japan about 120 km southwest of Tokyo (see Fig. 3.8). The impoundment site is located in a mountain-top depression with peripheral tailings dams located on three saddles as shown in Fig. 3.9. The site surficial geologic profile consists of volcanic tuffs bedrock with overburdens of varying degree of weathering. Construction of three tailings dams started in 1964. Weathered overburdens were stripped off by bulldozers to expose competent bedrock to form rough dam foundation surfaces. The Starter Dams were constructed of borrowed soils of volcanic origin. Damfill compaction was achieved by bulldozer traffics. Drainage conduits were installed on the dam bases to facilitate drainage of local springs. The cross section of 34 m high Dam No. 1 is shown in Fig. 3.10, while the section of 25 m high Dam No. 2 is shown in Fig. 3.11. Both dams were raised by the upstream construction method as shown in their cross-sections. The relatively lower Dam No. 3 was constructed as a conventional embankment dam with no dam raise.

1978 Izu-OhshimaKinkai Main and Aftershocks

On January 14, 1978, a magnitude 7.0 main shock occurred in the southeastern area offshore of Izu Peninsula. Figure 3.12 shows contour lines of estimated peak horizontal acceleration during the main shock in Izu Peninsula. The locations of the main shock and a series of aftershocks are shown in Fig. 3.13. As shown in the figure, the epicenters of aftershocks were moving westerly towards the Izu Peninsula. The largest aftershock was a magnitude 5.8 quake that taking place at 7:30 a.m. in the morning of January 15, 1978 followed by a magnitude 5.4 quake at 7:36 a.m. in proximity of the Hozukizawa tailings impoundment.

Dam No. 1 failed on January 14, 1978 due to liquefaction of the tailings during the main shock, while Dam No. 2 failed on January 15 about 24 h after the main shock and 5 h after the largest aftershock (see Figs. 3.14 and 3.15). The following

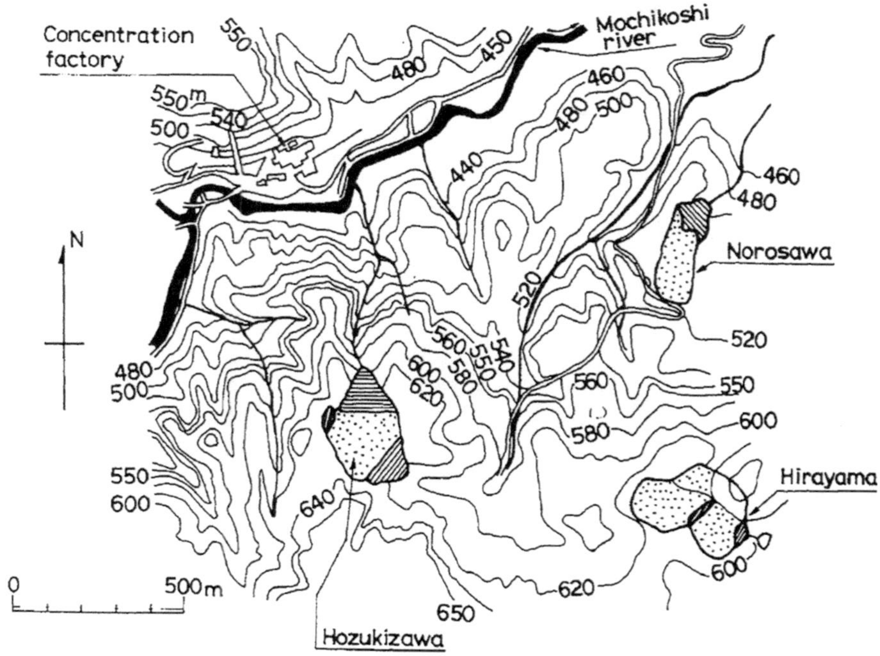

Fig. 3.8 Location of Mochikoshi concentration plant and Hozukizawa tailings impoundment (Ishihara, 1984)

descriptions of the respective failure of the two dams are based on eyewitnesses' accounts (see Ishihara, 1984).

Dam No. 1 failed almost at the same time as the major shakings of the main shock died down. A caretaker who stationed at a house on the left abutment immediately came out to check on the dam. He observed a bulged downstream dam slope. Tailings breach occurred in the upper portion of the dam segment near the left abutment. The ensuing rush of the debris flow mowed down trees and scoured valley slopes. As the flow reached its junction with the Mochikoshi river, it jumped over the opposite bank of the river channel and deposited debris on the road beside the channel. The debris flowed down the Mochikoshi river, and settled down on the river bed to a thickness of 1–2 m over a distance of 800 m. The flowing debris travelled further into the downstream Kano river and polluted a 10 km reach all the way down to the river mouth. As shown in Fig. 3.10, all the dam raises above the Starter Dam were gone with the tailings effluent. A total volume of 80,000 m^3 of solids escaped from the impoundment: three quarters were stored tailings together with one quarter of volcanic ashes used to build the retaining dykes above the Starter Dam.

At 8:30 a.m. the next day, one hour after the magnitude 5.8 aftershock, a monitoring crew discovered 5–6 relatively short longitudinal cracks about 1–3 m long and 5 mm wide on the downstream dam slope of Dam No. 2. At 9:30 a.m. another crew found a larger longitudinal crack, 5 m long and 5 cm wide in the middle of

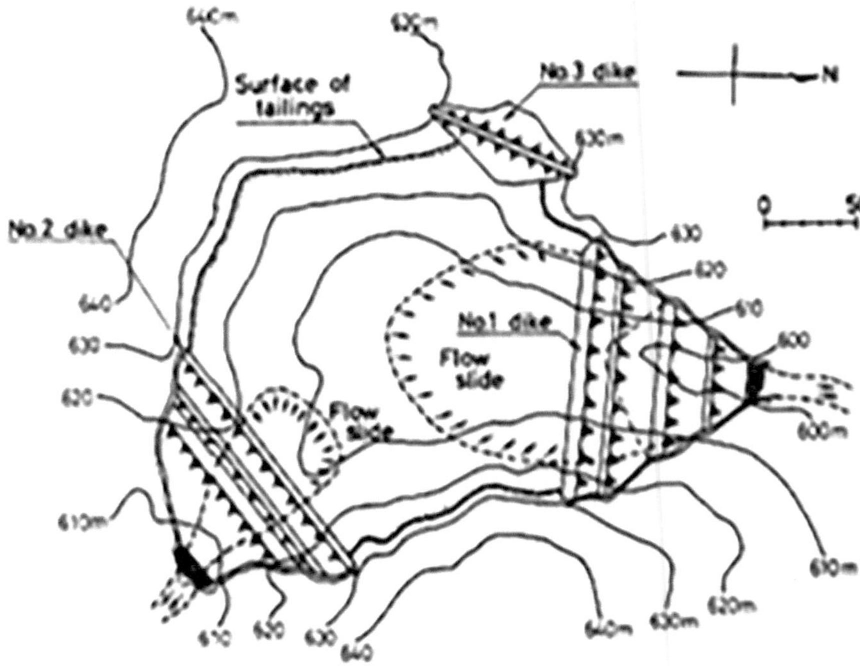

Fig. 3.9 Plan of Hozukizawa tailings impoundment and three peripheral tailings dams (Okusa & Anma, 1980)

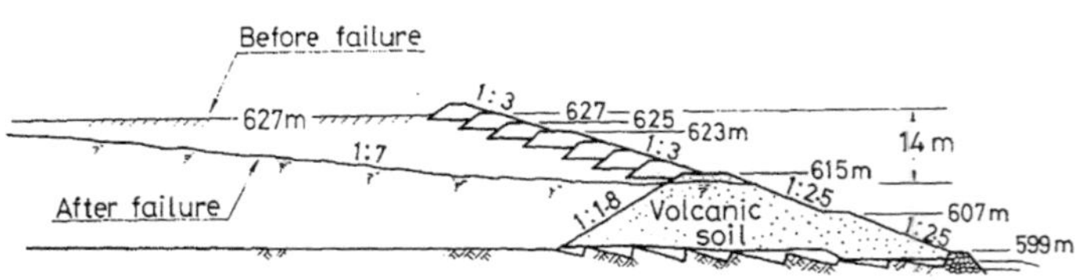

Fig. 3.10 Cross-section of Dam No. 1 (Ishihara, 1984)

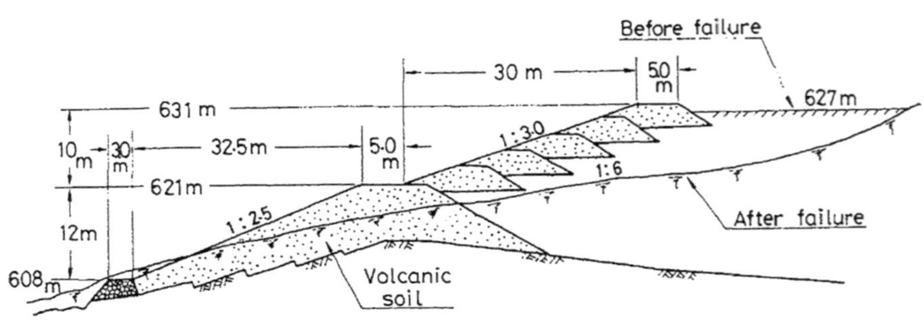

Fig. 3.11 Cross-section of Dam No. 2 (Ishihara, 1984)

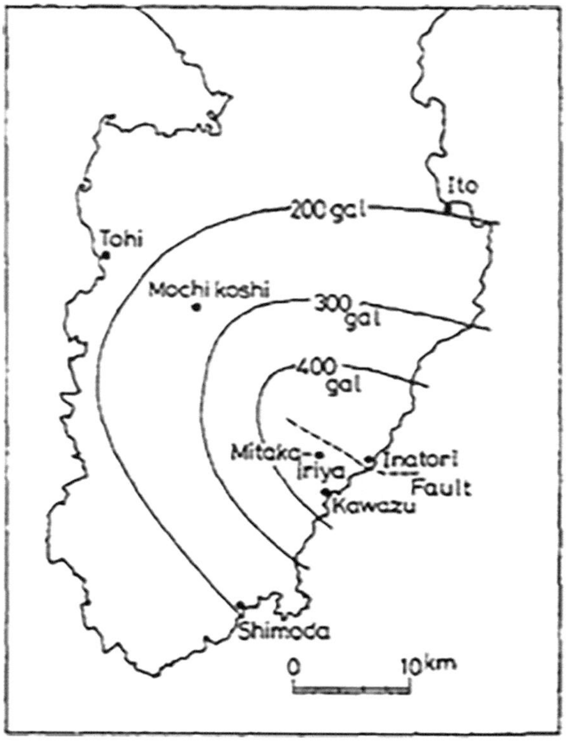

Fig. 3.12 Contours of estimated peak horizontal acceleration in Izu-Peninsula (Ishihara, 1984)

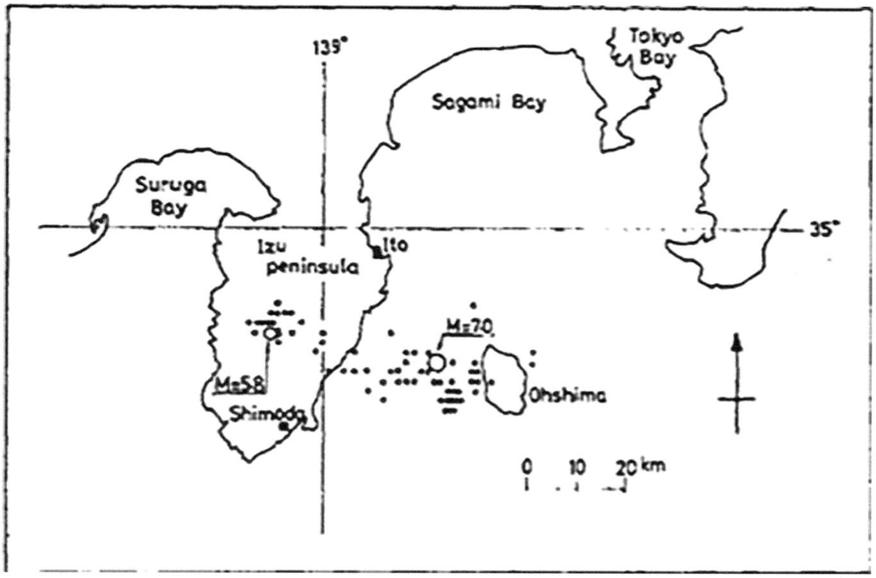

Fig. 3.13 Locations of main and aftershocks of Izu-Ohshima-Kinkai earthquake (Ishihara, 1984)

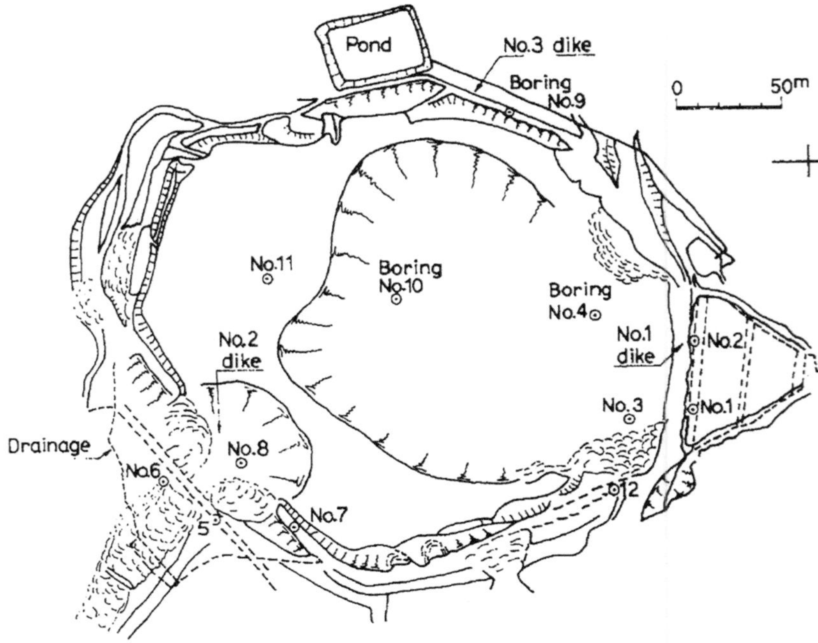

Fig. 3.14 Location plan of Hozukizawa tailings storage impoundment at Mochikoshi Gold Mine (Ishihara, 1984)

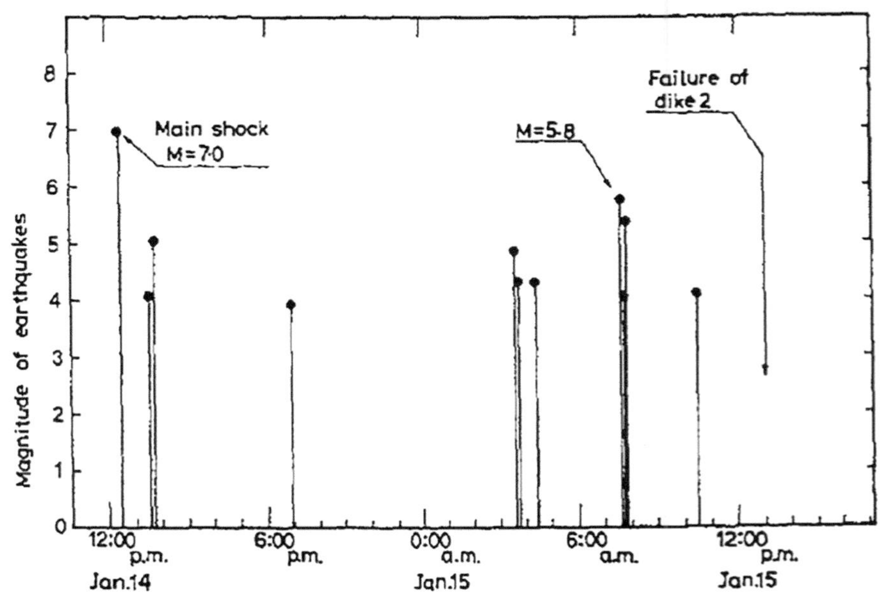

Fig. 3.15 Occurrence of aftershocks with magnitude > 4 following the main shock (Ishihara, 1984)

the downstream dam slope. At 1:00 p.m., a caretaker standing on the opposite side of Dam No. 2 noticed that the central segment of the dam was sinking gradually. Running to the dam, he observed that a 20 m long dam segment dropped suddenly resulting escapement of stored tailings from the impoundment. As the breach length grew to 65 m, a total volume of 3000 m^3 solids flowed away, including 2000 m^3 of stored tailings and 1000 m^3 of damfills, travelling for a distance of about 240 m. The breached Dams No. 1 and No. 2 as well as the back arcs of tailings slide scars left inside the tailings impoundment are shown in Fig. 3.14.

Post-earthquake Investigation

A post-earthquake investigation was set up by a Japanese Investigation Committee, who issued a report in 1979 regarding incidents of Mochikoshi Tailings Disposal Dams (Ishihara, 1984). A program of field and laboratory work including drilling, sampling, index and engineering property testing were carried out to better characterize the physical conditions of the foundation, tailings dams and stored tailings. Ishihara (1984) based on the database assembled by the committee carried out further static and seismic stability analyses to arrive at the following qualitative conclusions:

1. Dam No. 1 failed due to liquefaction of tailings towards the end of the main shock;
2. The delay of eventual failure of Dam No. 2 was due to progressive increase of excess pore pressures resulting from the main shock and a series of aftershocks (see Fig. 3.15) as well as redistribution of these excess pore pressures due to seepage (see Fig. 3.16).

Byrne and Seid-Karbasi (2003) carried out a numerical modelling of Dam No. 1 during the mainshock. Their finite-element numerical model is shown in Fig. 3.17. The calculated displacement of the dam at the end of the main shock is shown in Fig. 3.18. The lower portion of Fig. 3.18 shows the exaggerated grid distortion. Touileb (2013) provided an overview regarding dam design considerations with the respect of the largest aftershock.

Lessons Learned

The 1978 Hozukizawa Tailings Dams seismic liquefaction flow failure demonstrated following facts:

1. Tailings dam constructed by the upstream method is inherently fragile, especially for a location with significant seismic hazard;
2. A dam requires a robust design to resist a maximum considered earthquake appropriate to its seismo-tectonic setting. To achieve this goal, one has to consider the potential for a substantial aftershock occurring shortly after a main shock;
3. Post-earthquake investigation and analyses including the review of eyewitness' accounts of the earthquake impact enhance the value of the case history; and
4. Touileb (2013) provided a good overview of case histories involving dam safety and aftershocks.

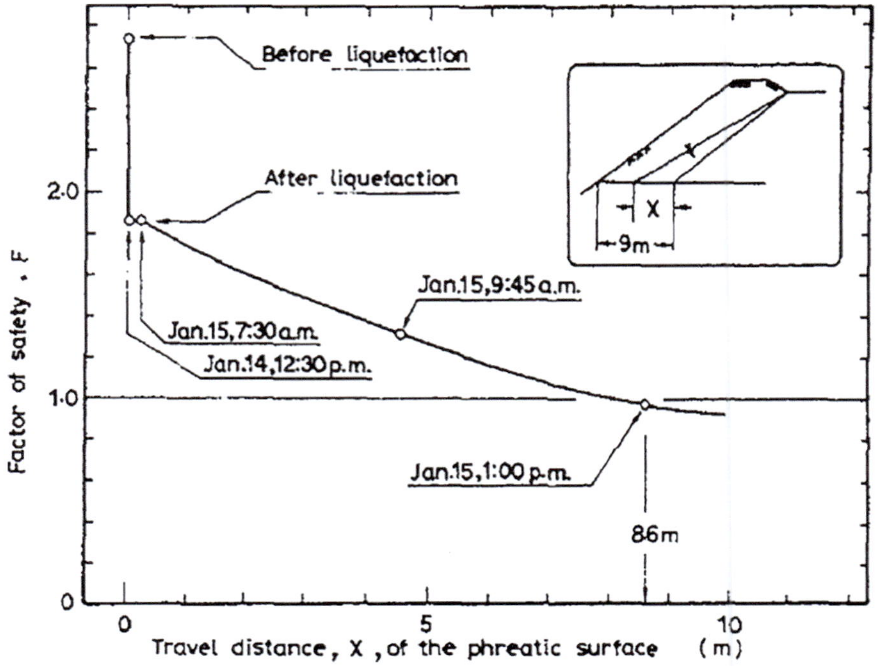

Fig. 3.16 Change of safety factor caused by movement of phreatic surface after tailings liquefaction (Ishihara, 1984)

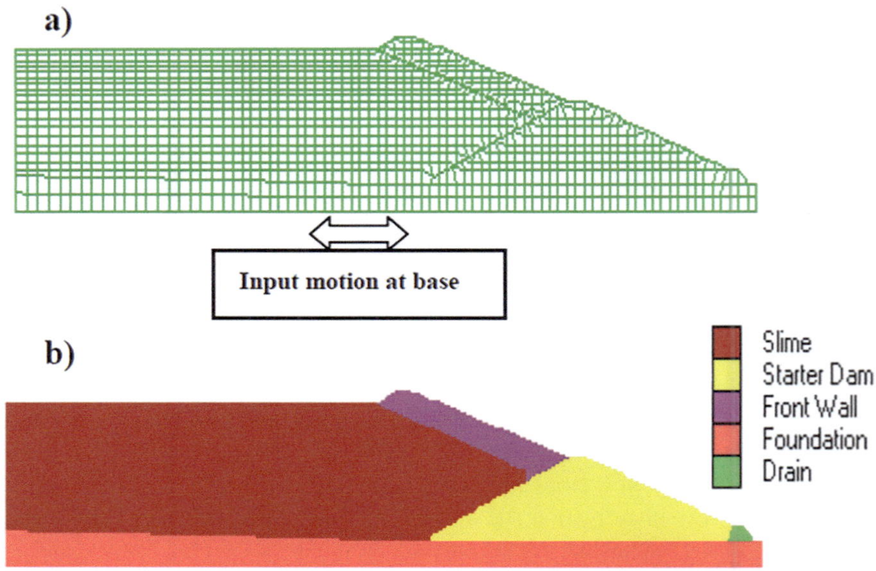

Fig. 3.17 Finite element numerical model of Dam No. 1 (from Byrne & Seid-Karbasi, 2003)

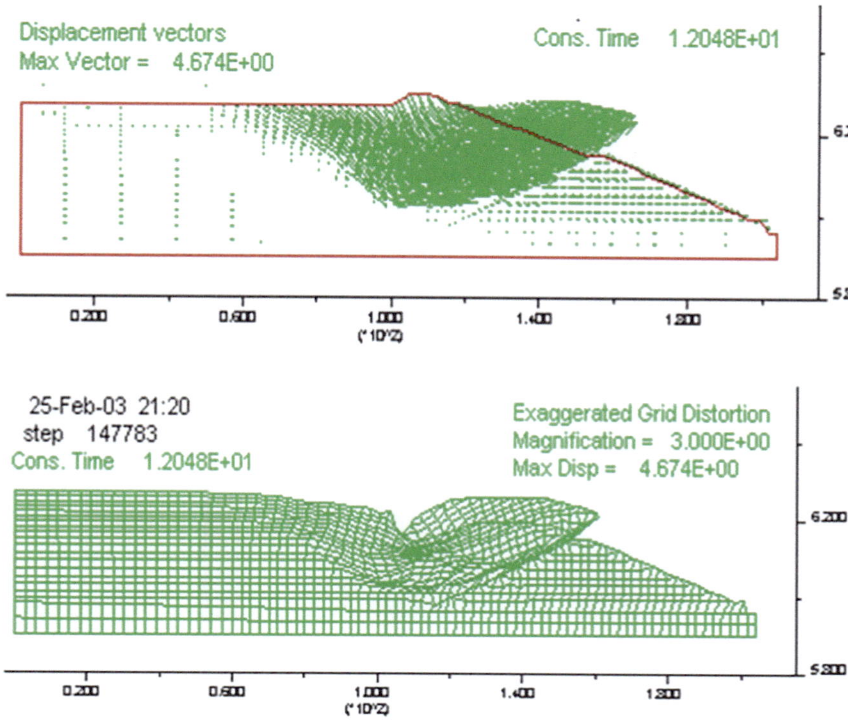

Fig. 3.18 Calculated displacement of Dam No. 1 (lower figure with exaggerated grid distortion, Byrne & Seid-Karbasi, 2003)

3.6 Case 5 December 16, 1983 Ok Tedi Tailings Dam Foundation-Excavation Triggered Landslide, Papua New Guinea

Ok Tedi gold and copper mine (OTML) is located at Mount Fubilan in the Star Mountains of Papua New Guinea. On December 16, 1983, a landslide involving approximately 35 million cubic metres of soil and rock was triggered by foundation excavation for the damsite of a new tailings storage facility. Protracted litigation related to the insurance claim regarding the landslide eventually reached the Supreme Court of Papua New Guinea in November 1989. The case was settled after one month of court proceedings, although details of the settlement were not made public (Griffiths et al., 2004).

This case is unusual for the following reasons:

- There was no dam failure. Rather, the chosen damsite was covered by massive landslide debris of soil and rock;
- The mine is located in Papua New Guinea, an island in the southwestern Pacific Ocean north of Australia. The region has high precipitation and seismicity in

complex seismo-tectonic, geologic and geomorphologic settings. Culturally, the land is owned by indigenous communities and administered according to their customs. The country's economy is in a transition period from a subsistence level through rapid development to a modern level; and

- Two decades after the initial landslide, finding an optimal tailings management scheme in Papua New Guinea remained a technical challenge to all stakeholders including the federal, provincial and local governments, general public and populations affected by the mining development as well as international institutions related to financial and technological support (Murray et al., 2003).

Ok Tedi Mine Operation

Figure 3.19 shows the location map of the Ok Tedi mine and its vicinity. The mine is located near the border between the West Papua Province of Indonesia and Papua New Guinea (PNG) as shown in the inset map. The geomorphological map of the landslide failures as well as the footprint and crest of the proposed dam and the location of Section Line A-A_1 are shown in Fig. 3.20. The engineering geological cross-section along Line A-A_1 through the damsite is given in Fig. 3.21.

The following description is mainly based on Griffiths et al. (2004). Feasibility studies started in 1976 established the mineral reserves at OTML as a leached capping layer of 88 million tonnes of ore with an average gold content of 1.75 g/tonne overlying on 290 million tonnes of sulphide copper porphyry with 0.59 g/tonne of gold and 7400 g/tonne of copper. Moreover, there were additional 33 million tonnes of copper porphyry deposits with 0.99 g/tonne of gold and 13,650 g/tonne of copper. To extract gold and copper from these ores, 1.1 billion tonnes of mine waste would have to be properly managed.

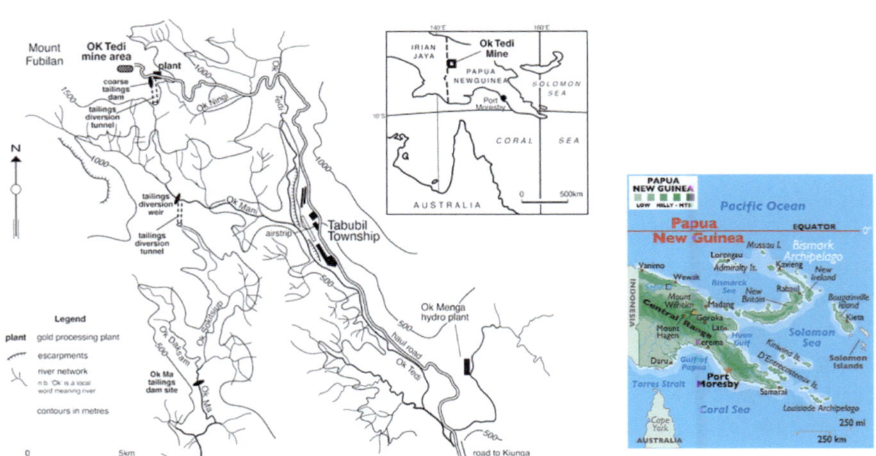

Fig. 3.19 Location map of Ok Tedi mine, its vicinity (from Griffiths et al., 2004) and Papua New Guinea inset

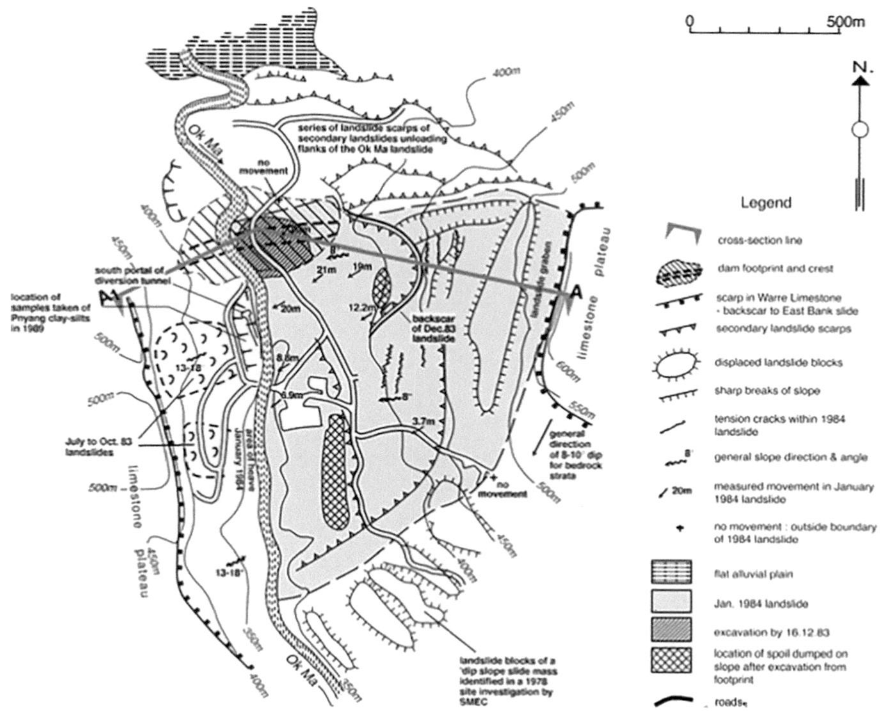

Fig. 3.20 Geomorphological map of the 1983/1984 damsite landslide failures (from Griffiths et al., 2004)

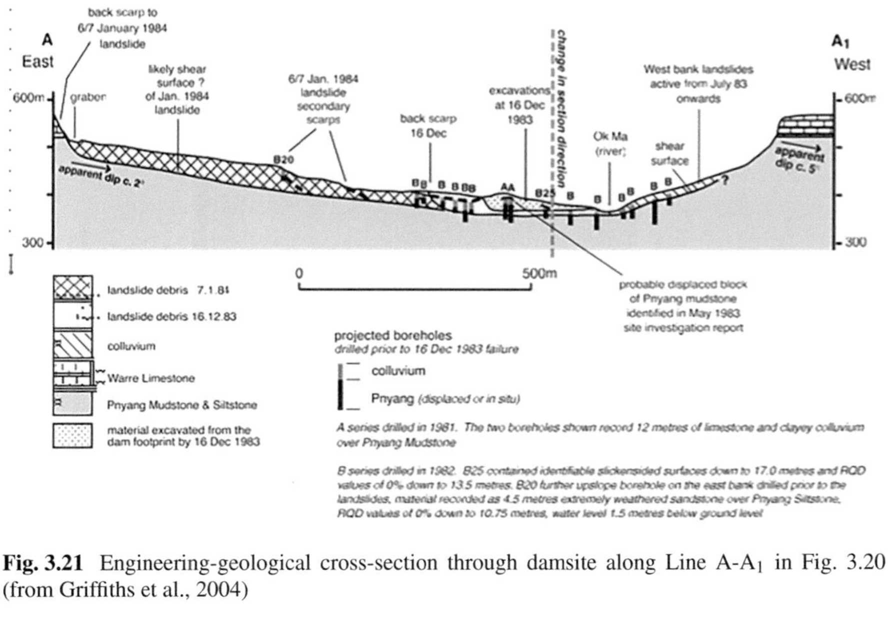

Fig. 3.21 Engineering-geological cross-section through damsite along Line A-A_1 in Fig. 3.20 (from Griffiths et al., 2004)

The mine development plan involved first mining the gold cap layer followed by mining the underlying copper ore. To take advantage of the high gold price at the time, the mine adopted a "fast track" approach. It started construction in late 1980 based on the feasibility study, and conducted the final design work at the same time with the target date of mid-1984 for commissioning the mine. Unfortunately, the site conditions deviated from design assumptions considerably. By 1982, the mine setbacks included the following:

1. A 140 km long haul road from the mine at Mount Fubilan to the port of Kiunga on the Fly River, via the newly built mining township of Tabubil, took seven months longer to build than expected. It was completed in July 1982 and was over budget, because of heavy rainfall, unstable terrain, high ground relief and lack of suitable borrow areas.
2. Unsuitable chosen plant sites.
3. The Ok Menga hydro-electric scheme was replaced by a diesel generation station due to difficult site conditions.

Ok Tedi Tailings Management

The 1983/1984 landslides triggered by excavations at the tailings damsite changed the course of tailings management scheme at Ok Tedi Mine. From the start, the government required that the mine built an on-land tailings storage facility before it began its mining operation. The foundation excavations started on November 4th, 1983. By December 5, 0.25 million tonnes of material had been excavated from the dam footprint area, and various spoil piles were stockpiled on the slopes above the east bank of the Ok Ma river (see Figs. 3.20 and 3.21). On December 16th, a landslide of about 3.4 million cubic metres from the eastern slopes moved about 5 m into the dam footprint area, and stopped all construction works there. During the night of January 6/7, 1984 a major enlargement of the landslide involving a total of 35 million cubic metres occurred. The sliding mass moved a maximum of 21 m downslope, essentially destroying the damsite. Soon after the enlarged landslide, the mine initiated the insurance claim.

While the lawsuit was going on, an interim scheme of tailings management was pursued by the mine with ongoing concessions by the government to keep the mining operation continuing. First, a small dam known as the ITS (Interim Tailings Disposal), was built at the mine site to collect coarse tailings. A centrifuge process was utilized to remove mine waste cyanide used for gold extraction. The fine tailings were discharged into the Ok Tedi River, down the Fly River with some perhaps reaching the Gulf of Papua (see Fig. 3.19). In October 1989, when the coarse tailings storage facility was full, the government further allowed the mine to discharge coarse tailings also into the river system.

This Interim Tailings Disposal Scheme (ITDS) continued until mid-1998. During this period, 30 million tonnes of mine waste was discharged into the Ok Tedi River every year. The estimated ecological cost included: fish stocks falling by 70–90%, and estimated loss of about 1350 km^2 of vegetation due to river sedimentation and flood over a period of 20 years. Litigation initiated in 1995 by the people of the

Fly River Basin against the mine was settled out of the court. The mine paid out nearly £300 million to resettle ten villages, to compensate 30,000 local landowners, to mount a pollution clean-up operation, and to pay the landowners' legal fees. A dredging project in the lower Ok Tedi region was initiated in 1998 to contain the impact of the tailings. It removed around 20 Mt of material annually, about half of this is tailings. The government, having 20% of the mining company's shares at that time, was balancing its revenues from the project against the accompanying ecological damages. Currently, the mine is owned by the PNG Government.

Four major impacts of this increased sediment load on the Ok Tedi and Fly River systems can be discerned (Banks, 2001).

1. First, both tailings and waste rock derived material are transported by Ok Tedi and Fly River systems (see Fig. 3.22) and, as a consequence, sediment levels in the river increased substantially. Fish biomass in the Ok Tedi River has been reduced by 90% over pre-mine levels, and in the Middle Fly River, the reduction is around 50%. Fish numbers and variety to date appear to be undiminished in the Fly River flood plain and side streams.
2. Second, the deposition of this sediment has a major impact. While up to a third of the sediment input, including waste rock and tailings, is carried through the system, there has been significant aggradation. The bulk of aggradation has

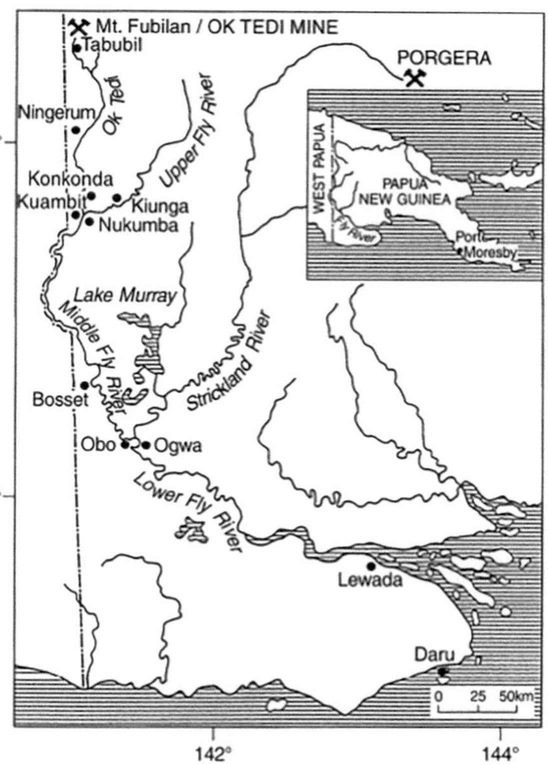

Fig. 3.22 The Fly River system—areas impacted by the Ok Tedi interim tailings management scheme since 1989 (from Hettler et al., 1997). Sampling in the Middle Fly River floodplain concentrated on the reach between Bosset and Obo

occurred below the break in slope between the more mountainous upper Ok Tedi River, and the flat terrain from the Lower Ok Tedi River downstream. This has resulted in a build-up of mine sediment in the Lower Ok Tedi River that has caused a rise in the river bed. Prior to the Bige dredge remediation project which commenced in the mid 1990s, the aggradation led to a smothering of vegetation and raised water tables resulting in "die-back" over an area of 500 km^2, with a further 200 km^2 of vegetation in the Upper Middle Fly River under stress. It is the extent and nature of this dieback that had the most profound impact on the ecology and human populations of the Lower Ok Tedi River area, and can be directly linked to the legal remedies sought by the affected communities. The obvious effects include destruction of riverbank gardens, loss of sago stands, significant forest resources, and biodiversity, and reduced navigability of rivers. Less obvious are the lifestyle and cultural impacts of the destruction of such a large area of forest, gardens and riverbank villages.

3. Third, there have been increasing dissolved copper concentrations recorded in the Lower Ok Tedi River and Middle Fly River since the early 1990s. Dissolved copper levels are of concern because copper can have a severe impact on flora (especially algae) and fauna at relatively low levels. To date, the annual average copper concentrations in the Fly River are well within PNG, Australian and WHO drinking water standards. However, monitoring has revealed elevated short-term spikes or pulses in the recorded levels. The origin and impact of these spikes are less well understood. The longer-term impacts of heavy metals in the ecosystem are less clear, and there is a potential for Acid Rock Drainage development from waste rock and tailings that accumulate.
4. Fourth, some concerns have been raised over the longer-term effects on the Fly River Delta of the deposition of elevated sediment levels. Concerns have been expressed over the impacts on commercial and subsistence fishery stocks, and over localized Acid Rock Drainage problems from deposition of tailings and waste rock material.

Review of Damsite Landslides

Griffiths et al. (2004) reviewed the causes of the construction-induced landslides at the damsite by first describing the complex geologic, geomorphologic settings of the site. The area has an average annual rainfall greater than 8 m and is covered in dense, primary and secondary rain forests (see Fig. 3.23). The bedrock geological sequence consists of:

(1) Tertiary Pnyang Formation Warre Limestone Member (40–100 m thick)—This forms a hard rock-cap above the underlying Pnyang Mudstone on the edges of the river valley. It is a cream, grey or buff coloured, algal and coralline limestone. The Warre limestone is described as having a conformable transgressional boundary with the underlying Pnyang mudstones and siltstones.
(2) Tertiary Pnyang Formation Mudstone Member (1000 m thick)—These deposits are exposed in the valley sides and river bed. It is a thinly to massively

Fig. 3.23 Typical terrain in the mine area (Griffiths et al., 2004)

bedded dark grey silty mudstone and clayey siltstone with calcareous beds. One calcareous marlstone bed provides a clear marker horizon in the sequence.

The Ok Ma valley lies on one limb of a low-angle pitching anticline. At the damsite the rocks are dipping at approximately 8–10° with a dip direction of 205–210°. Evidence of relatively low strength, polished bedding planes in the upper 2 m of the Pnyang mudstone was found in the field before the 1983 foundation excavation. The 1977 aerial photograph (Fig. 3.24) shows that the river has a straight course at the dam site with alluvial valley meanders developed upstream and incised meanders downstream. This geologic history had two important influences: (1) the Pnyang Mudstone Member contains existing slickensided shear surfaces, possibly developed due to tectonic folding; and (2) the chosen damsite is situated in the area of numerous past landslides.

Figure 3.25 shows correlations of residual shear strength with the plasticity index of Pnyang Formation clay-silts and clayey colluvium. Other index properties of these two soils are given in Tables 3.3 and 3.4. These data indicate that these soils have a wide range of varying properties. Without thorough understandings of detailed subsoil profiles including existing slickensided shear surfaces, it would be very difficult to have sufficient margins of safety to design and execute the tailings dam construction, especially without adequate time in the adopted "fast track" approach.

Back stability analyses of the 1983/1984 landslides showed that the residual shear strength along the failure surface is 10.7–13°, assuming that the prevailing water table is at the ground surface. However, the design of the damsite foundation excavations was based on the assumption of an average residual shear strength of 21°. This assumption of a critical geotechnical parameter was not corrected in time to prevent the occurrence of the 1983 and 1984 progressive landslides that ruined the damsite. The mine operation system at the time was not set up to allow timely feedback of realistic field conditions to check and modify design assumptions until the ongoing dam construction was rudely stopped by the massive landslides.

Griffiths et al. (2004) summarized their review of the landslides occurrence by examining eight prerequisite criteria (see Table 3.5) listed by Peck (1969a) for successfully applying the "Observational Method" in carrying out geotechnical

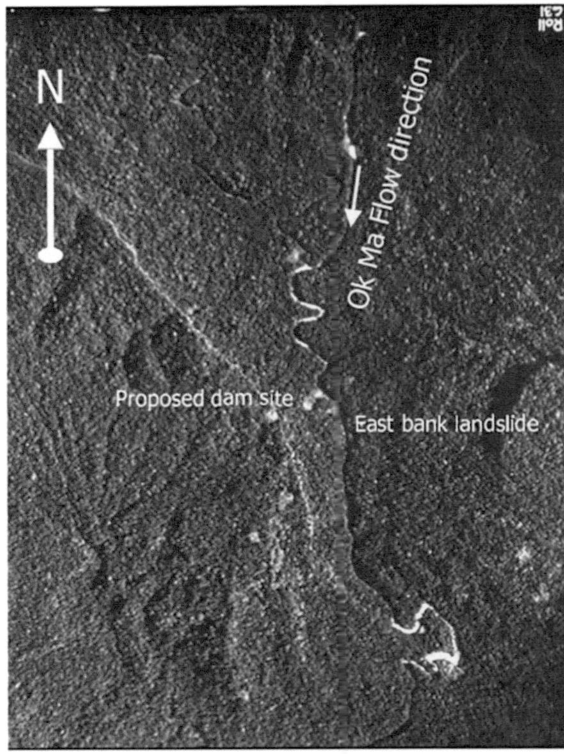

Fig. 3.24 1977 aerial photograph of the damsite (from Griffiths et al., 2004)

projects. In their estimation, the original design likely met only the first, but overlooked the rest seven criteria. In other words, they carried out their original design without the iterative feedback loops to check and improve their design based on what they learned from ongoing focused observations during construction.

On-Land Mine Tailings Disposals in Papua New Guinea

Leaving aside the specific case history of the Ok Tedi mine damsite 1983/1984 landslides, it is useful to learn lessons from the collective experiences of other mines regarding their tailings management methods adopted at Papua New Guinea. Murray et al. (2003) reviewed the obstacles to on-land mine tailings disposal in Papua New Guinea. They took the approach of reviewing the impact of the following factors on the full life cycle of the tailings management system from the construction, operation to closure phases:

- Climate;
- Tectonics, Topography and Geomorphology;
- Geology, Geotechnique and Geochemistry;
- Social and Flora/Fauna; and
- Constructability and Logistics.

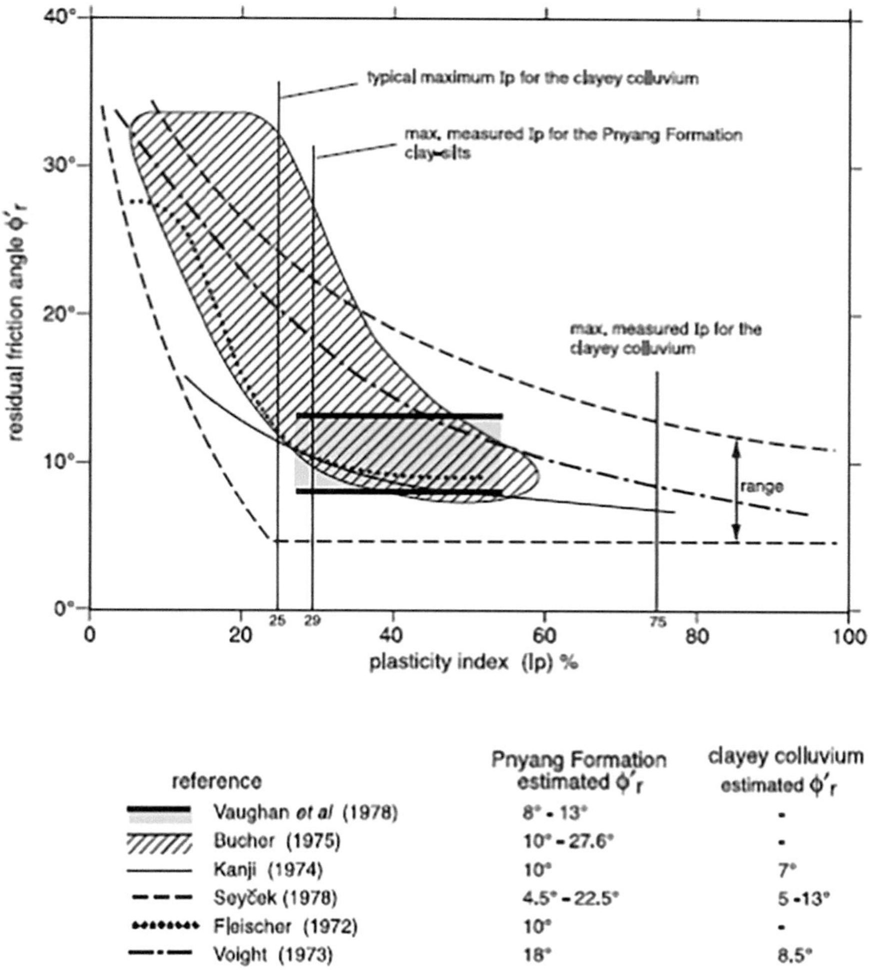

Fig. 3.25 Shear strength of Pnyang Formation clay-silts and clayey colluvium (from Griffiths et al., 2004)

Table 3.3 Typical values of index properties for the Pnyang clay-silts and clayey colluvium (from Griffiths et al., 2004)

Index property	Pnyang clay-silt	Clayey colluvium
Moisture content (%)	10–25	7–102
Plastic limit (%)	14–25	5–75
Liquid limit (%)	18–50	21–108 (typically 34–50)
Plasticity index (%)	1–29	3–75 (typically 16—25)
Clay fraction (< 2 μm) (%)	4–42	6–44
Undrained shear strength (kPa)	*c.* 25	*c.* 19

Table 3.4 Index properties and residual shear strength values for undisturbed samples of Pnyang clay-silts taken from area of flexural shearing due to tectonic folding (from Griffiths et al., 2004)

Sample number	Liquid limit (%)	Plastic limit (%)	Clay fraction (< 2 μm) (%)	φ_r' (with c_r'—0) in °
Block 1	38	23	35	19 (natural shear)
Block 2	39	23	37	17 (cut plane)

Note Locations, where samples Block 1 and Block 2 were taken in the 1989 post-landslides investigation, are shown in Fig. 3.20

Table 3.5 Important features of the "Observational Method" (from Griffiths et al., 2004)

1. Sufficient exploration to establish at least the general nature, pattern, and properties of the deposits, but not necessarily the detail
2. Assessment of the most probable conditions and the most unfavourable conceivable deviations from these conditions. In this assessment, geology often plays a major role
3. Establishment of the design based on a working hypothesis of behaviour anticipated under the most probable conditions
4. Selection of quantities to be observed as construction proceeds and calculation of their anticipated values on the basis of the working hypothesis
5. Calculation of values of the same quantities under the most unfavourable conditions compatible with the available data
6. Selection in advance of a course of action or modification of design for every foreseeable significant deviation of the observational findings from those predicted on the basis of the working hypothesis
7. Measurement of quantities to be observed and an on-going evaluation of the actual conditions
8. Modification of design to suit the actual conditions

Climate

The climate of PNG is generally warm tropical with temperatures ranging from 20 to 30 °C, with only minor diurnal variations. This makes for a pleasant climate at about 1000 m elevation, but it can be oppressively hot and humid on the lower coastal plains. At higher elevations, frosts often occur during July and August. For on-land tailings storage facilities, important climate factors are (1) normal rainfall, (2) wet-period events and (3) dry-period events:

(1) Normal rainfall—The tropical climate of PNG is affected by several climatic components including the tropical convergence zone, where vertical transfer of heat energy occurs and is released through violent thunderstorms. The climate can be divided into the following two distinct seasons:

- The south-east (trade wind) dry/drier season is usually from June to September; and
- The north-west (monsoon) wet/wetter season is usually from October to May.

(2) Wet-period events—The wet season occurs during the monsoon season. However, rainfall is often characterised by storm cells with high spatial variability, even for relatively small catchments. Parts of PNG such as the Ok Tedi mine site are among the wettest in the world, with annual rainfall well exceeding 10 m per year, with little or no dry season. In contrast to the Ok Tedi area, Porgera, Misima and Lihir receive a more modest 4–5 m of rainfall annually. Many of the islands located south of about 5° South, such as Misima, are located in the tropical cyclone belt. These areas can be hit by storms resulting in rainfalls of over 1 m in 24 h.

Although annual rainfall totals are high by world standards, the long return-period events, which typically control designs of dam and water diversion structures, tend to be lower than those in regions of higher latitudes. Another significant feature in rainfall patterns is that the range in rainfall event magnitude for different return periods is much smaller in PNG than those in higher latitudes. This means that diversions/spillways or other similar surface water management structures are regularly tested at or close to design capacity. Therefore, there is little room for error or neglect. This situation is especially relevant for closure design. It implies that water management structures will be tested many times at or near design capacity and a high level of maintenance will be needed after mine closure.

(3) Dry-period events—Many areas of PNG have a distinct dry season. In the Wau area near the Hidden Valley mine, it has an average of about 1.8 m annual rainfall. Of which only 0.4 m falls during the predominantly dry season in the four months from June to September. El Nińo refers to the atmospheric/oceanic phenomenon that occurs in the Equatorial Pacific, resulting in a rainfall deficit over much of Australasia including PNG. El Nińo events occur every five to seven years. A potential negative impact on on-land tailings storage from the dry-period droughts is desaturation of tailings with subsequent exposure to oxidation, resulting in the potential for acid generation. However, extensive studies at Ok Tedi show that the annual rainfall throughout the mainland PNG may be sufficient, even in drought years, to provide saturation of tailings without the need to maintain a supernatant pond.

Tectonics, Topography and Geomorphology

Earthquake activities across PNG are among the highest in the world. The seismicity of New Guinea is dominated by the oblique collision zone between the margins of the Australian Plate and the Pacific Plate. The lateral extension of the plate boundaries is illustrated in plan view in Fig. 3.26, and in cross-section in Fig. 3.27. The Pacific plate is moving west at a rate of 68 mm per year, while the Australian plate is moving north easterly at ~ 20 mm per year. Within the convergence zone of these two major plates lie four micro-plates: the Woodlark, Solomon and the North and South Bismark plates. Margins of these micro-plates define the volcanic and seismically active Island arc areas of New Guinea.

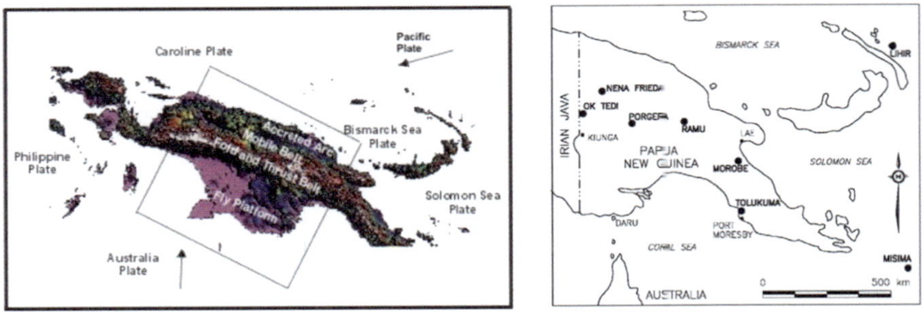

(a) Papua New Guinea Main Island and adjacent islands (b) Papua New Guinea and Ok Tedi Mine

Fig. 3.26 Oblique collision of the Australia and Pacific plates (from Murray et al., 2003). **a** Papua New Guinea Main Island and adjacent islands. **b** Papua New Guinea and Ok Tedi mine

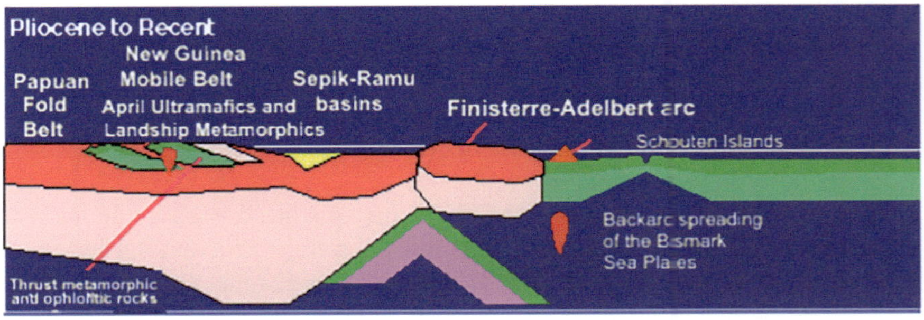

Fig. 3.27 Cross section of plate boundaries in the collision zone adjacent to New Guinea Mobile Belt (from Murray et al., 2003)

In the last 30 million years, the Australian plate margin along the north coast of New Guinea has been mostly under compression. In central New Guinea such as the Ramu-Sepik basin, the area has experienced episodes of extensional dilation as the plate motions shifted. The heavy compression loads along the margin of the Australian plate have created onlapping thrust faults and folds in the northwest/southeast trending Papuan Fold Belt, where the central highlands and mountains around Ok Tedi, Porgera and Nena/Frieda are located. This compression also resulted in a thickening of the crust and mantle which raised the oceanic crust at the north margin of the Australian plate. This thicker and more stable raised area has become the Fly River Platform. To accommodate the strain due to the uneven compressional loads across the region, a series of NNE trending faults have formed.

PNG Main Island has steep and rugged mountainous terrain, often combined with near-vertical karstic limestone cliffs in the elevated regions. This mountainous terrain is characterized by steep, deeply incised, river valleys, which are subject to flash flooding, high erosion rates and valley wall instability. Typically, the major rivers and tributaries have bed gradients in the order of 5–10%. The rivers usually have a sediment-carrying capacity exceeding the sediment load. Hence many streams are actively down cutting, and flow in steep-sided (> 30° side slopes) valleys. This topography generally limits the available capacity for tailings storage.

Dense vegetation cover also tends to obscure the terrain. Trees grow higher in the valleys and lower on the ridges. Visually, the vegetation cover smooths the real ground topography. Low visibility and lack of tracks and clearings add to difficulties to ground reconnaissance of potential tailings storage sites. Applications of modern airborne geophysics and LiDar methods have improved the collection of reliable/economical terrain data.

There are extensive areas of relatively flat coastal plains across PNG. These plains tend to be remote from the ore bodies. The southern coastal plain, through which the Fly River discharges, is an area that has slowly flooded due to sea level rises after the last ice age. The area, known as the Fly River platform, is part of the stable Australian tectonic plate and has relatively low seismicity. The Bige site where Ok Tedi stores dredged sand waste is on the north margin of the southern coastal plain at about 20 m above sea level, and over 400 km from the ocean. Sedimentation studies on the Fly River flood plain suggest that historic flood plain deposition only averages a few mm per 100 years. This is orders of magnitude less than the current rate of rise of the mountainous highland due to tectonism.

The north coastal plain is strongly influenced by active tectonism and uplift. The Sepik and Ramu flood plains, for example, are characterized by extensive sediment build-up with very recent lacustrine/marine/alluvial deposits filling valleys right up to the emerging foothills. This creates broad flat-bottomed valleys flanked by ridges that are geometrically well suited for a potential tailings impoundment. Problems are relatively high seismicity and foundation materials of very low shear strength and/or potentially liquefiable.

Typical geomorphologic hazards for a potential tailings storage site include the following:

- Landslides and mudflows—both regional structurally controlled slides in weak rock and localized colluvium slides in wet mountainous tropical terrain. Typical slides in coarse and fine colluvium are shown in Fig. 3.28. The fine colluvium slide as shown in Fig. 3.28b involved over 2 Mm^3 of slide material blocking temporarily a major river.

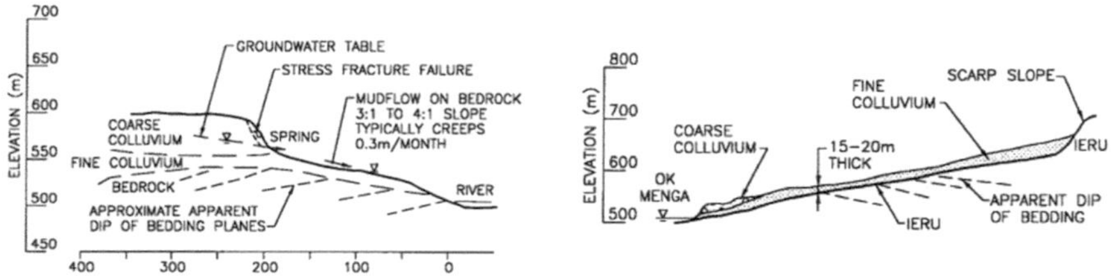

(a) Coarse colluvium landslide (b) Fine colluvium landslide

Fig. 3.28 Typical cross-sections of coarse and fine colluviums over Pnyang mudstone bedrock (from Murray et al., 2003). **a** Coarse colluvium landslide. **b** Fine colluvium landslide

- River-blocking landslide-dams—floods caused by the uncontrolled breach of landslide-dams triggered by earthquakes are well documented in PNG.
- Debris flows—severe and continued debris flows can reduce impoundment storage capacity and cause wave activity in the impoundment.
- Rock avalanches—rock avalanches in PNG can travel tens of kilometres as fluidized masses of sand, gravel and boulders. A large historic rock avalanche occurred over 8000 years ago on the Tabubil plateau at the Ok Tedi mine, involving ~ 7000 Mm^3 of sliding materials.

Geology, Geotechnique and Geochemistry

All PNG rock types tend to be at the low end of the shear strength range for their family, likely due to the weathering process in a tropical humid setting. Granodiorite becomes little more than soils with sugary texture; schist being highly graphitic; mudstones and shales containing extensive shear zones at very low residual friction angles. Deep tropical weathering profiles or unconsolidated quaternary alluvial deposits are common in areas with flat topography not subject to high erosion or slope instability. High rainfall and acids from vegetation help create Karstic terrains in the central highlands limestone belt. Relevant aspects of typical rock and soil profiles are outlined below:

- Sedimentary Rocks

Weak sedimentary rocks include very poorly indurated marine and freshwater shale and sandstone sequences such as the Ieru (cretaceous), Pnyang (miocene) and Chim (cretaceous) formations. These sediments can have unconfined compressive strengths of around 5–10 MPa when fresh, and disintegrate (slake) rapidly upon exposure, forming extensive muddy colluvium deposits. Typical friction shear strength along weak bedding planes can be 12° or less. Thus, if an unfavourably dipping rock structure is present, even very flat ground slopes can be unstable. Any storage facility design and site selection need to address the following conditions:

- slope underlain by out-dipping shales can be unstable at angles as flat as 6°;
- large rafts of bedrock can be displaced by landslides for many hundreds of metres; and
- excavation for spillways or other purposes can slake and regress rapidly, unless protected from weathering.

- Volcanic and Intrusive Rocks

Deep weathering can also reduce the strength of volcanic and intrusive rocks. Deep weathering can reduce these rocks to residual soils of weak, sugary texture. Examples of this weathering are found in the Frieda River area.

- Karst Features

Weak limestone, high runoff water and acids derived from rotting vegetation produce extensive karst features in PNG. These features include vadose (above groundwater level) and phreatic (below groundwater level) caves, crevice karst and solution sinkholes. The proposed Lukwi tailings damsite at Ok Tedi mine (Klohn et al., 1987) to replace the damsite damaged by the foundation-excavation-triggered landslides includes the following:

- Crevice karst developed along prevalent joint sets—The crevices vary from a few metres to over 40 m deep and from a few centimetres to over 10 m wide. The crevice karst develops in exposed limestone, but quickly reduces toward the valley abutments if the limestone is covered by insoluble, low permeability rocks such as shales.
- Dolines or sinkholes can develop if collapses of underground cavities propagate to the ground surface. Underground caves can form at the intersection of crevice karst. Dolines many kilometres in diameter are not uncommon. Smaller sinkholes to a few metres in diameter can be very difficult to detect, especially if masked by colluvium cover.
- Vadose cave development is controlled by the direction and gradient of free or perched water flow, joint orientation, bedding planes or layers of solution-resistant rock. In an active cave, there is minimal sediment infill. In abandoned flow paths, the voids can be quickly infilled with sediment.
- Phreatic caves formed below previous or existing water tables are often oval-shaped, and infilled with highly plastic residual clay or sand sediments.
- Karst conditions require careful evaluation regarding water tightness of the tailings dam and integrity of the tailings impoundment.

• Weathered Tropical Soils

Weathered soils in PNG include saprolite, laterite, limonite and andosols, etc. They are formed by insitu chemical weathering and alteration of rocks in tropical climates. The resulting soil profiles are less consistent than those formed by sedimentary processes. Weathered soil profiles are problematic due to their unexpected and highly variable geo-mechanical behaviour. These soils can have extremely high moisture content to the point that free water can often be squeezed out by hand pressure. The high water content is obscured by the apparent stiffness of the soils upon exposure. The stiffness is due to weak cementation at the contact points of the platy lattice soil structure. Once disturbed, these soils can collapse and flow. A surprising feature is the widespread occurrence of sinkholes in the residual soil terrain. These sinkholes seem to be formed by the erosion of water running beneath a hardened often limonitic crust. The water erodes the underlying softer soil until the surficial caprock is undermined and collapsed. These features are both insidious and difficult to detect.

• Colluvium

Colluvium characteristics are closely related to the parent rock type. They can often be grouped into two principal types: coarse (bouldery) colluvium and fine (silty) colluvium (see Fig. 3.28). Coarse colluvium is derived from the collapse of the mountain top, avalanche or similar catastrophic failure. It is a relatively more competent material. Re-cementation of coarse colluvium, particularly for those with limestone content, also can occur. Fine colluvium, on the other hand, is developed by weathering processes from sedimentary or volcanic rocks. It has the properties of silt or clay with very low undrained shear strength. Its effective-stress friction angle rarely exceeds 20°. Low residual strengths (< 12°) are common at the interface of the fine colluvium and the bedrock surface. Colluvium characteristics grade between the two main types, depending on whether the material properties are dominated by the fine matrix or the coarse fraction. Often landslide activity can lead to mixture of fine and coarse colluvium, resulting in high heterogeneity of material properties.

Geochemistry—Geochemical issues have not been a major negative factor in the development of mines in mainland PNG. In the highlands, extensive limestone occurs within or adjacent to the ore bodies, forming a large buffering capacity. Most streams flowing from the PNG highlands have a natural pH of at least 8. The wet climate allows many options to keep any acid-generating material saturated. However, seismic-induced tailings dam deformation and erosion from the high rainfall need to be properly addressed from a geochemical perspective.

Social and Flora/Fauna

The social and community issues within PNG are complex and critically important to the feasibility of an on-land tailings storage facility. Land ownership and unencumbered access to it are fundamental to local populations. Temporary tenures of mining operations and their long-term environmental impacts including those resulting from tailings dam failures can have detrimental social impacts. This was the case for one scheme investigated for the Ok Tedi mine. Tailings storage schemes for large projects can occupy the total land area for a particular clan. Whether the scheme occupies part or all of a clan's land, the necessary displacement of people from the project area is fraught with difficulty and may not be practicable because most of PNG is held in customary ownership, and permission may not be readily attained to temporarily relocate people onto another clan's land. Overlapping claims of ownership along area boundaries create an additional problem. This may be resolvable, given sufficient time and with sincere negotiations. Equitable compensation to all parties that claiming ownership may be required. Compensation aside, the long-term impact of a mining operation on the local economy and community life and subsequent loss of incomes from mining employment after mine closure need to be carefully planned and appropriately managed to prevent the detrimental consequences of mining-induced boom-and-bust economic cycles. Other factors related to the closure of tailings impoundment are: appropriate control of any potential geochemical problems of acid generation and/or metal leaching. Sustainable mining development requires the protection of local species of flora and fauna. Thus, pre-project flora and fauna studies need to be carried out adequately to demonstrate that species sustainability will not be compromised. Strong vegetation growth is a benefit in rehabilitating of the mining properties in preparation for its closure. However, negative effects of uncontrolled vegetation growth and/or cumulations of dead woods and branches need to be prevented to ensure proper functions of the hydraulic structures such as ditches and spillways.

Constructability and Logistics

Issues regarding the constructability and logistics of building tailings storage facilities in PNG are outlined below:

- Site Access and Preparation Works—The construction of site access roads involves many challenges such as the disposal of excess cut spoil, potentials for re-mobilization of small and large-scale landslides, stream and runoff management, availability of suitable pavement materials, etc.
- Construction Materials—Availability of suitable construction materials is a significant issue, because most materials at shallow depth contain high moisture content and low-strength, normally-consolidated materials. Lengths of good weather days and hauling distances impact the efficiency and economics of earthwork productivities.
- Management of Water Erosion and Sediment—The use of best construction practices could mitigate but not eliminate problems associated with erosion and sedimentation.
- Equipment Availability—Local contractors are engaged to maximise the socio-economic benefit of the mining projects to the local regions and PNG as a whole. However, the availability of suitable types and numbers of local construction equipment is a serious issue due to poor equipment maintenance regimes. This problem is particularly acute during the initial preparation and final closure phases of a specific mining project.

Summary of PNG Mining Projects

PNG has numerous challenges as indicated above for tailings management. These factors make the construction, operation and closure of on-land tailings storage facilities extremely difficult and expensive. Therefore, as summarized in Table 3.6, among 10 mining properties in the table, only Morobe and Nena Frieda chose on-land tailings storage schemes. However, even in 2022, the Nena Frieda scheme is still at the feasibility study stage. The smaller Morobe scheme, which started construction in June 2007, seems to encounter some uncertainty due to the difference between the provincial and federal government https://www.thenational.com.pg/clash-over-project/. It appears that the provincial government prefers to continue the work on the existing on-land tailings storage facility (see Fig. 3.29, Mudd et al., 2020), while the federal government is in favour of a Deep Sea Tailings Placement. Thus, the eventual tailings management schemes to be adopted by Morobe and Nena Frieda remain to be determined.

Banks (2001), in his Papua New Guinea Baseline Study for the International Institute for Environment and Development and World Business Council for Sustainable Development, pointed out the following two areas should be targeted in terms of policy reform:

- Fiscal regime reform—the issue of intergenerational transfer of resources should be looked at, with a view to re-establishing a stabilization and/or investment trust fund for at least a portion of current government mineral receipts; and

Table 3.6 Summary of PNG mining projects[a] (from Murray et al., 2003)

Property	Status	Mineral	Tailings (tpa)	Tailings total mine (t)	Disposal method
Bougainville	Closed	Copper/gold	31,000,000	> 500,000,000	Riverine
Ok Tedi	Operating	Copper/gold	29,000,000	> 600,000,000	Riverine
Porgera	Operating	Gold	6,500,000	> 100,000,000	Riverine
Tolukuma	Operating	Gold	110,000	≅ 1,500,000	Riverine
Misima	Operating	Gold/silver	5,000,000	> 60,000,000	DSTP
Lihir	Operating	Gold	3,800,000	> 80,000,000	DSTP
Ramu	Prospect	Nickel/cobalt	4,900,000	100,000,000	DSTP
Morobe	Prospect	Gold	4,900,000	> 37,000,000	On-land
Nena Frieda	Prospect	Copper/gold	70,000,000	> 800,000,000	On-land
Freeport[a]	Operating	Copper/gold	73,000,000	> 1,000,000,000	Riverine

[a] The Freeport mine is located within the Indonesian Province of West Papua, which forms the west region of the island New Guinea and has been included for completeness

[b] Tailings disposal methods include: Riverine, Deep Sea Tailings Placement (DSTP, Marine) and On-land

[c] Locations of mine properties are shown in Fig. 3.26b, except Bouganville in Fig. 3.19 coloured inset map and Freeport as indicated in Note a

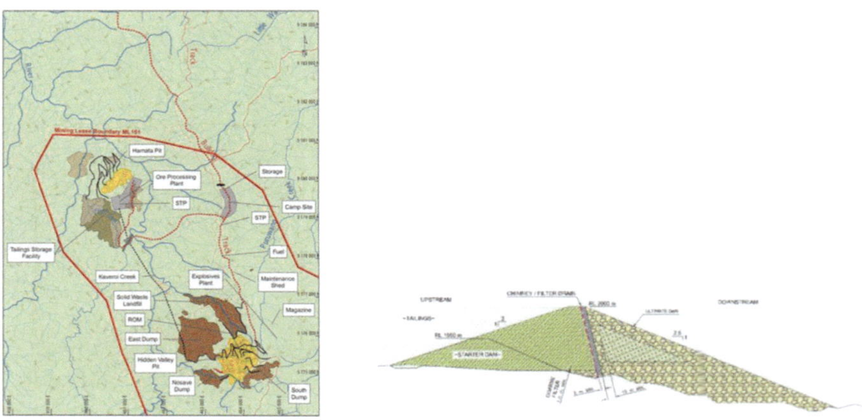

(a) Mine site plan (b) Main Tailings Dam cross-section

Fig. 3.29 Morobe Hidden Valley on-land tailings storage facility (from Mudd et al., 2020). **a** Mine site plan. **b** Main tailings dam cross-section

- Sustainability policy development—there is a need to tie the Sustainability Policy in with Environmental Policy, and broader National development issues. The Sustainability Policy should be concerned with mechanisms for benefit distribution (including intergenerational transfer of benefits at the local scale), mechanisms for enhancing sustainable use of benefits at local levels, and local-level governance in mining areas. The policy development process may also contribute to the development of mechanisms to ensure more effective and sustainable use of mining revenues at the national levels, particularly in terms of intergenerational transfer of revenues.

The above recommendations remain to be appropriate at the present.

3.7 Case 6 July 19, 1985 Stava Tailings Impoundments Overtopping Failure, Italy

On July 19, 1985 at Stava near Tesero (in Trento Region, northern Italy), two tailings impoundments located in Pozzle collapsed suddenly at 12.23 p.m. local time, causing the death of 268 people, the release of 180,000 m^3 of tailings, the destruction of many buildings and bridges and a total damage over 133 million Euros (see Figs. 3.30, 3.31 and 3.32).

Stava Tailings Impoundments

Since the sixteenth century, a mine was active on Mt. Prestavel for the extraction of small amounts of fluorite and, to a lesser extent, silver galena. In 1961, a new mill plant was constructed using the froth flotation system to obtain high-grade fluorite suitable for use in the chemical industry. To store tailings byproduct produced from the mill,

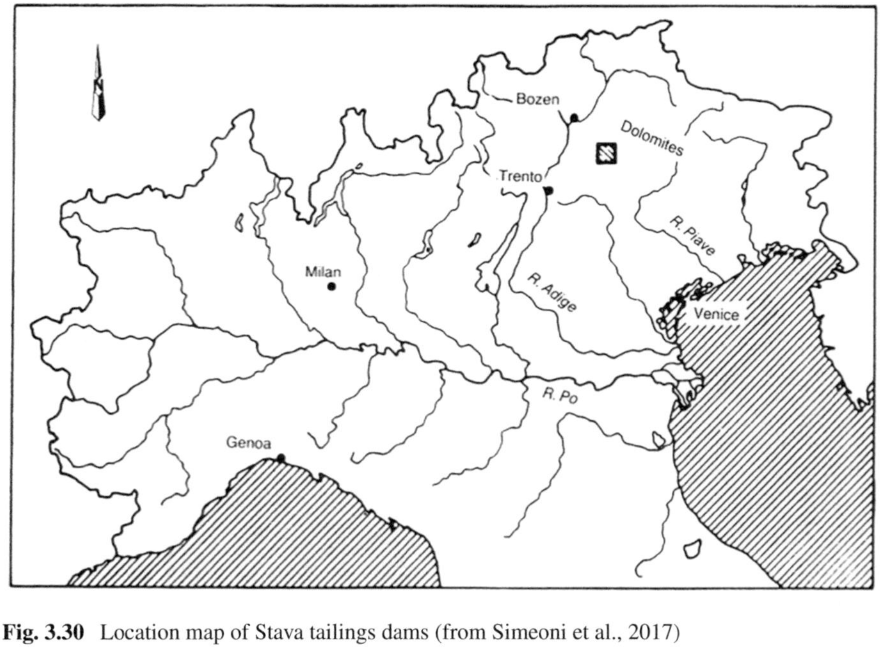

Fig. 3.30 Location map of Stava tailings dams (from Simeoni et al., 2017)

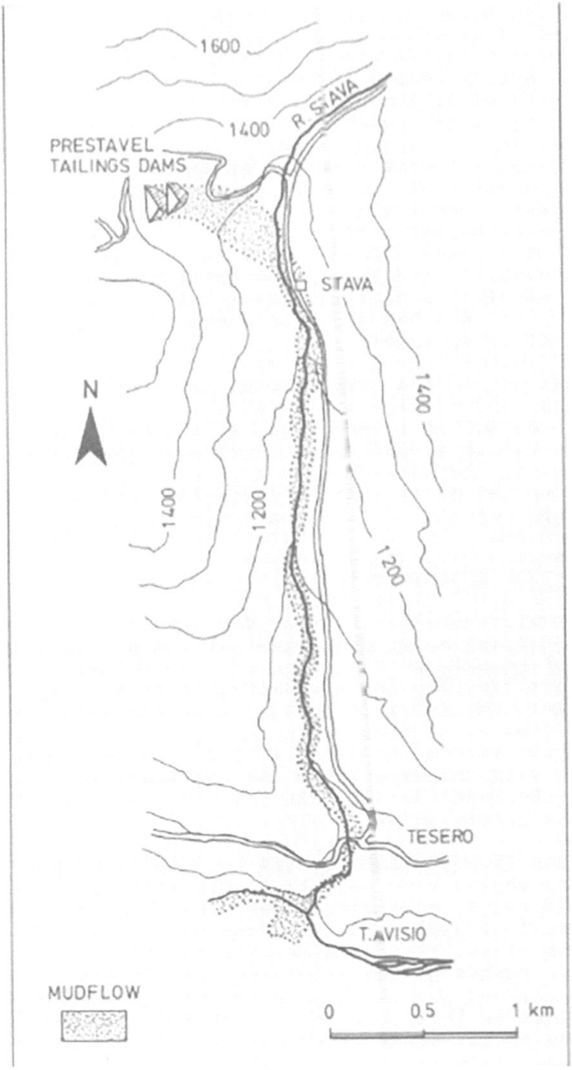

Fig. 3.31 Location map of the dam-breach mudflow travel course (from Berti et al., 1988)

two tailings impoundments were constructed at Pozzle to be retained respectively by a lower (25 m final height, constructed from 1961 to 1969) and upper tailings dams (34 m final height, constructed from 1969 to 1985 when dam failure occurred). Figure 3.33 shows the geological map of the area. The Pozzle tailings impoundments complex is illustrated in a transverse section in Fig. 3.34, and in a plan view in Fig. 3.35. Figure 3.36 shows a frontal aerial view of the contiguous impoundments taken in 1983.

As shown in Fig. 3.33, the Prestavel Mine is located in the Lagorai porphyry outcrop. Further south are the Werfern formation outcrops of alternating layers of

Fig. 3.32 Lower Stava tailings impoundment in April 1985 (from Simeoni et al., 2017)

mainly calcareous rocks and siltstones. The Quaternary fluvio-glacial deposits cover the tailings impoundments area. The Porcellini river is a surface stream uphill of the impoundments site. It disappears at the impoundments site and re-surfaces as a spring further downstream. As shown in Figs. 3.34 and 3.35, both tailings dams were constructed using the upstream method. While the two Starter Dams were constructed using local gravel borrow materials, the tailings dams were constructed using cycloned sandfill as dam construction material. Two moveable hydrocyclones were used to raise the Lower Tailings Dam, while one fixed cyclone was used for the Upper Tailings Dam. For each impoundment supernatant water collected in a pond at the far side was decanted through concrete drainage pipes which were installed along the impoundment base and extended through the base of the peripheral dam (see Figs. 3.35 and 3.37).

Decant concrete drainage pipes installed on the impoundment base apparently deformed under the overlying tailings weight and caused leakage of water and tailings from both impoundments: (1) in January 1985 at the Upper impoundment and (2) in May 1985 at the Lower impoundment with formation of sinkholes as shown in Fig. 3.35. Moreover, the precipitation season in the region is from May to August, and the precipitation in 1985 is unusually high in the months of May and June. Thus, the groundwater table was relatively high in the Pozzle tailings impoundments towards mid-July.

July 19, 1985 Cascade Collapse of Two Tailings Impoundments

On July 19, 1985 the mudflow caused by the cascade collapse of the Pozzle tailings impoundments flowed down the narrow Porcellini river valley on the west side slope of the River Stava (see Fig. 3.31) destroying and burying low-lying areas in the valley as shown in Figs. 3.38, 3.39, 3.40 and 3.41. Within a few minutes the frontal mudflow waves reached the small town of Tesero near the confluence of the Stava River and Avisio River about 4 km downstream from the impoundments site.

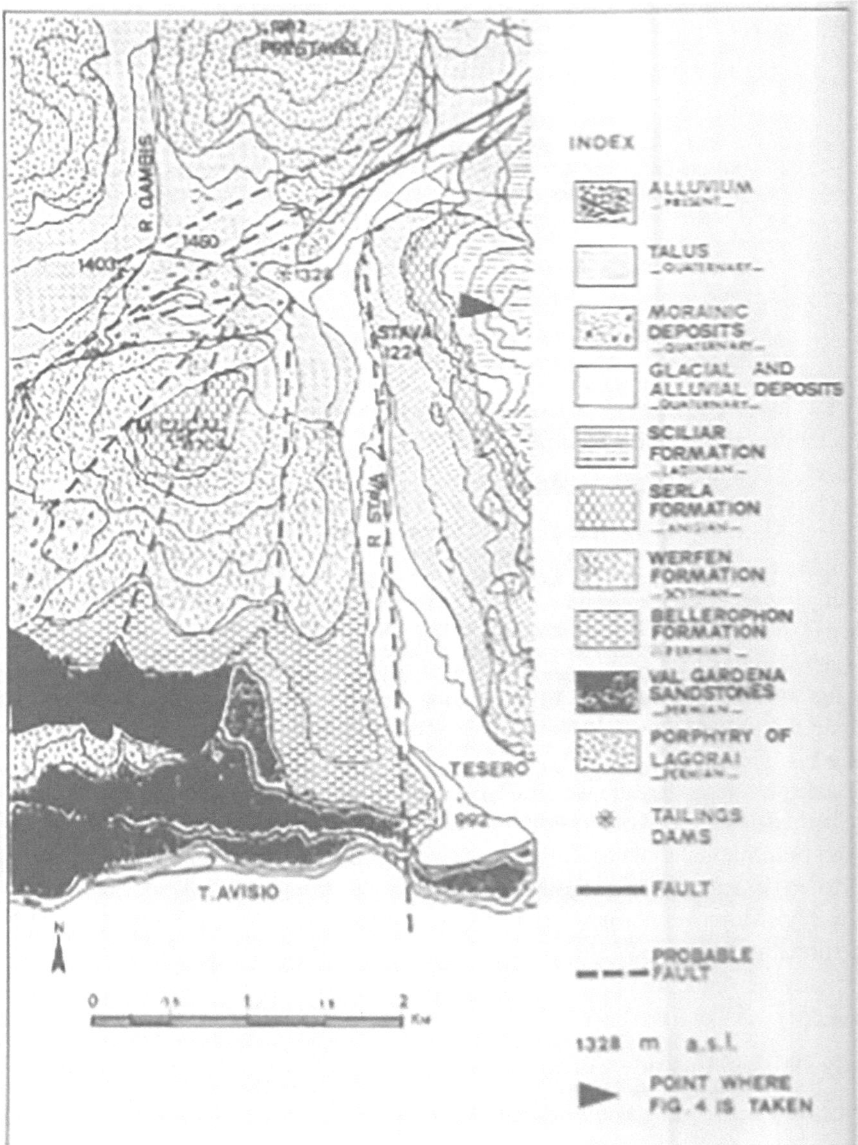

Fig. 3.33 Geological map of the Stava Valley (from Berti et al., 1988)

Redeeming Developments

Redeeming community activities started in the tragic afternoon of July 19, 1985 in organizing and executing the rescue operation. Thereafter the "Stava 1985 Foundation" charity was established to keep the memory of the 268 victims alive, to carry out educational activities and to promote further improvement of tailings dam

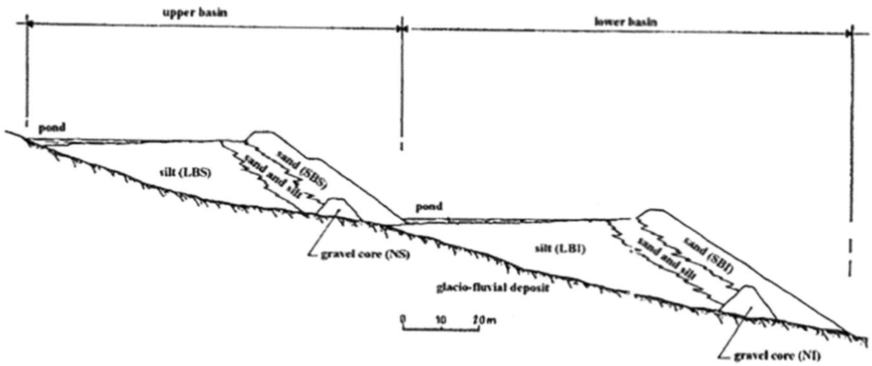

Fig. 3.34 Section sketch of the Stava tailings dams (from Colombo & Colleselli, 2003)

engineering (www.stava1985.it). Figures 3.42, 3.43, 3.44 and 3.45 show the manifestation of these redeeming developments. Figure 3.46 shows a recent satellite image of the Stava Valley.

Lessons Learned

The Stava impoundments failure is a tragic case that a tailings storage facility constructed by the upstream method managed to exist precariously for 25 years in a sensitive tourist region in the northern Alpine Italy. Its eventual sudden collapse was caused by the cumulation of a number of issues including technical defects, poor management and slack regulatory supervision. The dam breach exacted a heavy toll in terms of fatality and injury, economic loss and environmental degradation.

Positive aspects in response to this disaster are self-organized community rescue operation during the dam failure and the formation of the "Stava 1985 Foundation" charity thereafter to keep the memory of the 268 victims alive. The foundation has carried out educational activities and promoted further improvements of tailings dam engineering (see www.stava1985.it).

3.8 Case 7 October 17, 1990 Matachewan Tailings Dam Overtopping Failure, Canada

Baker et al. (1996) provided an overview of this case history, which illustrates how a costly tailings dam failure was developed insidiously over post-closure years due to the most critical flaw, neglect. Matachewan tailings impoundments were located 1.5 km west of the town of Matachewan, Ontario. The impoundments were used for tailings storage by Matachewan Consolidated Mines approximately from 1933 to 1953, initially as a gold mine and later as a custom milling operation for other mines in the vicinity. The impoundments were constructed by the upstream method using tailings. After 1953, the facility ceased operation with no provision for care and

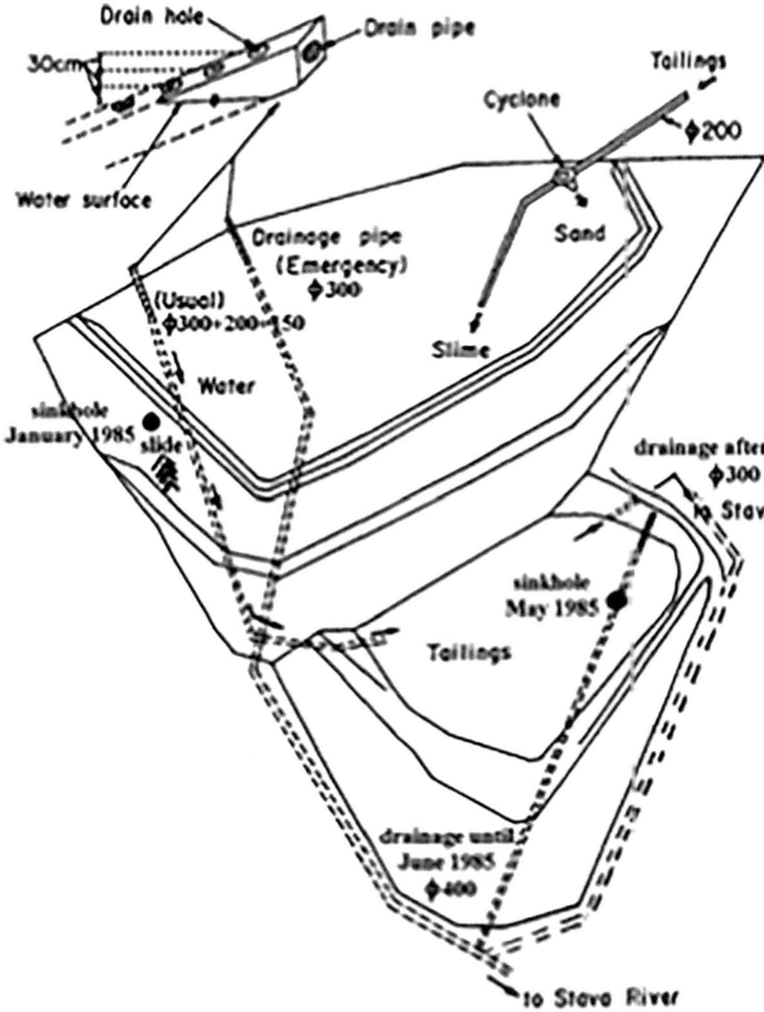

Fig. 3.35 Schematic plan of the decant pipes through Stava dams (from Colombo & Colleselli, 2003)

maintenance. The east portion of the tailings facility blocked a natural creek, which served as the original drainage outlet of Otisse Lake (see Fig. 3.47). For developing the tailings facility, the Otisse Lake drainage was diverted through an excavated channel to Hollinger Lake to the east.

Over the years the outlet of Otisse Lake was blocked by a beaver dam, which grew in size and raised the level of Otisse Lake by over 2 m. On October 17, 1990, after an extremely wet fall and several days of heavy precipitation, the rising water appears to have overtopped and breached a section of the south perimeter dam of the main tailings impoundment. The flowing water eroded the dam at the breached

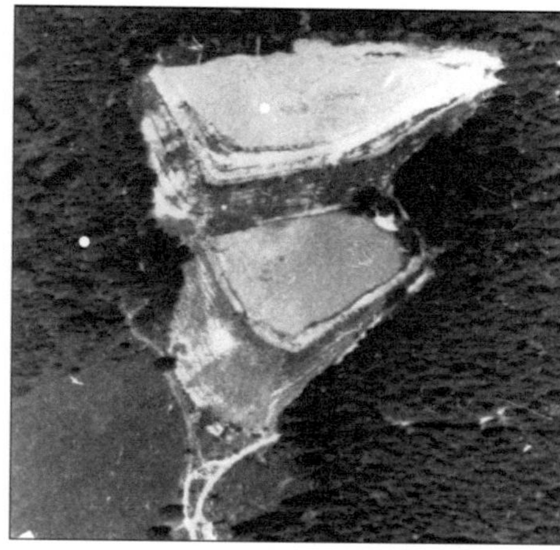

Fig. 3.36 Aerial view of Pozzle tailings impoundments in summer of 1983 (from Luino & De Graff, 2012)

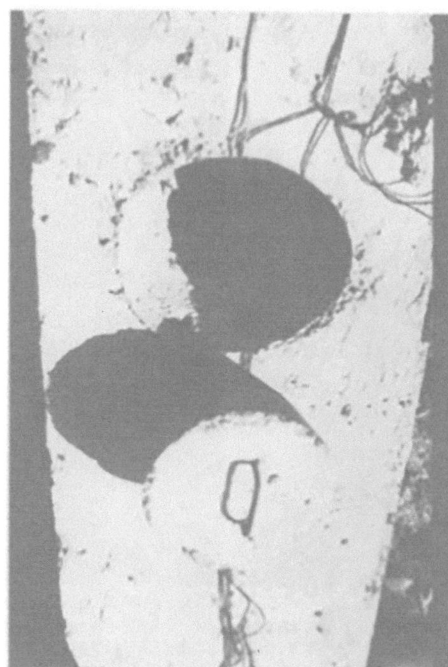

Fig. 3.37 Intake of the concrete decant pipe with a concrete plug (from Berti et al., 1988)

section and released about 195,000 m^3 of tailings material. A 1.5 km long section of highway was washed out. An estimated 55,000 m^3 of tailings were deposited in Davidson Creek downstream over a 1.6 km long reach to the confluence of the creek with the Montreal River. The tailings delta formed at the confluence damaged a beach and boat launch. Four private properties were also affected by the tailings deposits. There was no fatality in this incident.

Fig. 3.38 Looking towards Stava Village after mudflow swept through the valley (from Luino & De Graff, 2012)

Fig. 3.39 Before and after the passage of mudflow in the Tesero Village (postcard Trettel, P.) (from Luino & De Graff, 2012)

Fig. 3.40 Destruction of Tesero Village by mudflow (from Luino & De Graff, 2012)

Fig. 3.41 At confluence of Rivers Stava and Avisio (from Luino & De Graff, 2012)

Fig. 3.42 The monument to the Stava victims in Tesero cemetery (from Luino & De Graff, 2012)

Fig. 3.43 The Information Center of the Stava 1985 Foundation (from Luino & De Graff, 2012)

Fig. 3.44 Houses and hotels in the new village of Stava (early 2000s) (from Luino & De Graff, 2012)

Fig. 3.45 The Stava 1985 footpath on Mount Prestavèl (www.stava1985.it)

The emergency response and remedial measures were organized and coordinated by the Ontario Ministry of Natural Resources with the participation of 6 other ministries. Emergency response consisted of:

- Establishment of the response command centre and a media response team to facilitate effective management and communication;
- Construction of a "filter dam" to control the movement of tailings from the impoundment and a cofferdam to stop further inflow from Otisse Lake to the impoundment (see Fig. 3.47);
- Reconstruction of the washed out segment of Highway 566 to permit local access and trapping, and removal of tailings along the outflow paths upstream of the highway.

Remedial measures implemented subsequently included:

- Regrading the failure scarps in the main impoundment and vegetating the tailings surface;
- Improving the diversion of Otisse Lake drainage to Hollinger Lake to meet the probable maximum flood criteria;

Fig. 3.46 Recent satellite imagery of Stava Valley (from Google Earth Pro)

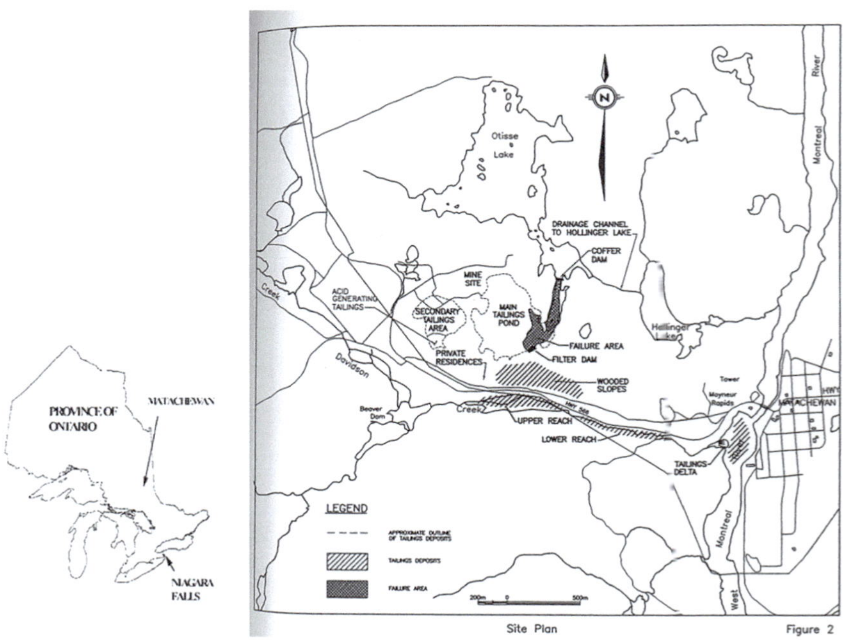

Fig. 3.47 Map of Matachewan tailings impoundments and vicinity (from Baker et al., 1996)

Fig. 3.48 Satellite imagery of Matachewan tailings impoundments and vicinity (from Google Earth Pro)

- Diverting runoff from the secondary impoundment and installing a surface cover to mitigate acid generation of the stored tailings;
- Restoring the drainage channel downstream of the main tailings impoundment, and providing a sand and gravel cover over the exposed tailings to prevent erosion; and
- Dredging the tailings accumulated at the mouth of Davidson Creek and returning the removed materials back into the tailings impoundment.

The total cost of the above remedial measures, excluding the emergency repairs immediately after the failure, exceeded $0.5 million. Only $44,500, about one tenth of this amount, was borne by the mining companies involved due to their limited financial resources. A recent satellite image of Matachewan Tailings Impoundments and its vicinity is shown in Fig. 3.48.

Lessons Learned

The overtopping dam failure was caused by inadequate closure of the mining operation. There were no care and maintenance and periodical inspections by the mine and regulators. However, the Ministry of Natural Resources responded to the situation promptly, and efficiently cleaned up spilled tailings and restored impoundment integrity.

3.9 Case 8 February 22, 1994 Merriespruit No. 4 Tailings Dam—Overtopping Failure, South Africa

The Merriespruit No. 4 tailings impoundment is located at the Harmony Gold Mine, Virginia, Orange Free State, South Africa. The peripheral dam was built using the conventional "paddock" deposition system practiced extensively in South Africa's gold mining industry. The dam was essentially raised by the upstream method and reached 31 m high, using a series of perimeter paddocks rising in elevation towards the upstream direction. The topmost paddock at the crest received spigotted tailings to form the "day wall", while excess decanted water drained to the upstream tailings beach. At night, for safety reason all tailings slurry was discharged into the central regions of the impoundment, with free water accumulated in the supernatant pond. Typical thickness of tailings deposits formed on the day wall ranged from 100 to 150 mm. The material was allowed to dry under the sun for about two weeks, while tailings deposition was moved around the perimeter. The dried material was then excavated to retrieve dam fill for constructing the next raise of the paddock wall (Wagener, 1997).

On February 22, 1994, following a 30–50 mm rainfall, the rising pond water overtopped a segment of the perimeter day wall. The dam had earlier problems prior to the major failure. Small slope failures had caused the impoundment to close temporarily and only mine water with small amounts of tailings were deposited. However, the deposition of these tailings caused the supernatant pond to shift its location from the centre to the perimeter of the impoundment against the dam (see Figs. 3.49 and 3.50). This shift of pond location was later revealed in a series of satellite images. As the pond shifted to the peripheral dam, it rendered the pond decant facility inoperable. Moreover, the gradient of seepage through the peripheral dam increased with the decreasing tailings beach width, and reduced dam stability.

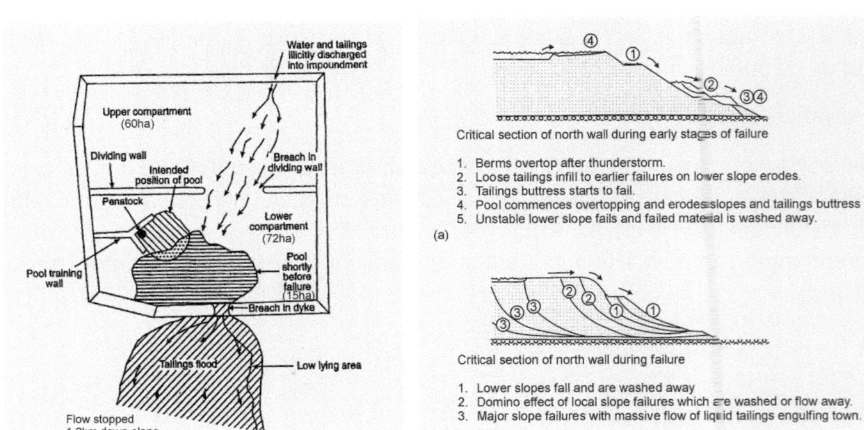

Fig. 3.49 Plan and section views of Merriespruit Dam failure (from Blight & Fourie, 2003)

Fig. 3.50 Breach of north perimeter tailings dam (from Minerals Council, 2020)

It was estimated that about 90,000 m^3 of pond water and 600,000 m^3 of tailings were released. The dam appeared to have failed through a process involving erosion, slope slump and washout with the slope slump scarps receding towards upstream until the dam was breached (see Fig. 3.49). The semi-fluid mixture of water and tailings flowed about 1.2 km and covered a 500,000 m^2 area causing severe damage to the village located in its path. The flood killed 17 people, and buried houses with many up to their roofs (see Figs. 3.51 and 3.52). The South African government appointed the Council of Scientific and Industrial Research to investigate this incident. As a result, the South African 1995 Draft Code of Practice for the "Design, Operation, and Closure of Tailings Dams" was introduced. Figure 3.53 shows a recent satellite imagery of Merriespruit and its vicinity.

Lessons Learned

The Merriespruit No. 4 Tailings Dam appeared to have failed through a process involving erosion, slope slump and washout with the slope slump scarps receding towards upstream until the dam was breached (see Fig. 3.49). The semi-fluid mixture of water and tailings flowed about 1.2 km and covered a 500,000 m^2 area causing severe damage to the village located in its path. The flood killed 17 people, and buried houses. As a result, the South African 1995 Draft Code of Practice for the "Design, Operation, and Closure of Tailings Dams" was introduced.

3.10 Case 9 August 19, 1995 Omai No. 1 Tailings Impoundment Spill, Guayana

Omai Gold Mines Ltd. was a joint venture of Montreal-based Cambior Inc. (65% share) and Denver, Colorado-based Golden Star Resources (30%) with Guyana's government (5%). The mine lies deep in the humid tropical Guyanese interior, 160 km from Georgetown, the capital (see Figs. 3.54 and 3.55). The mine started its open-pit mining operation in 1993 using cyanide extraction in a conventional carbon-in-pulp

Fig. 3.51 Merriespruit community inundated by tailings effluent (from Minerals Council, 2020)

Fig. 3.52 Flood inundation of downstream village (from Fourie et al., 2001)

process with a processing capacity of 13,000 t/day of gold-bearing ore. Finely-ground tailings slimes, predominantly minus 200 mesh size particles and residual cyanide solutions were stored in a tailings impoundment retained by No 1 Tailings Dam of 45 m height. Both the tailings impoundment and the mine itself lie on the banks of the Omai River (see Fig. 3.65). The Omai carries a flow of 4.5 m^3/s for a short distance before joining the Essequibo, one of the principal rivers of South America with a mean annual flow of 2100 m^3/s (Vick, 1996).

Fig. 3.53 Recent satellite imagery of Merriespruit and vicinity (from Google Earth Pro)

Fig. 3.54 Guyana and its location in South America (from Poole, 2008)

Omai No. 1 Tailings Dam

Figure 3.56 shows a cross section of Omai No. 1 Tailings Dam which consists of the following key damfill zones:

Fig. 3.55 Location of Omai mine and its vicinity (from Poole, 2008)

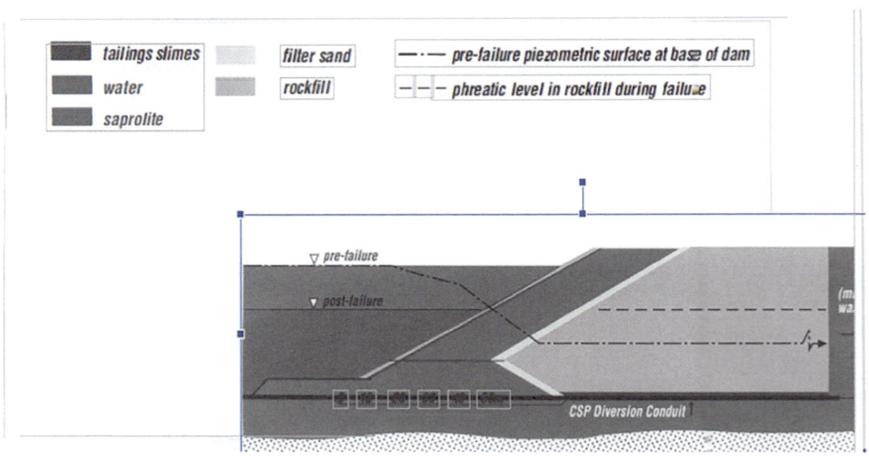

Fig. 3.56 Cross section of Omai No. 1 tailings dam (from Vick, 1996)

- a sloping impervious core of low-permeability, clayey saprolite (weathered insitu from bedrocks) soils;
- a thin zone of coarse-rock "riprap" along the upstream side of the core to protect it from erosion by wave and surface runoff;
- a primary downstream zone of compacted rockfill as the main body of the dam;
- a thin filter zone of sand to prevent migration of fine soil particles in the core to the downstream rockfill; and

- an auxiliary downstream zone of uncompacted mine waste (this zone is absent at both ends of the dam adjacent to the abutments.

Grain-size curves for various damfill zones are shown on Fig. 3.57. The saprolite core and rockfill materials are broadly graded. However, the filter sand is narrowly graded and is inadequate to serve filtering action and to prevent fine core material from entering the downstream rockfill zone (see Fig. 3.60). To facilitate the construction of the Starter Dam in dry conditions, a 900 mm diameter diversion corrugated-steel-pipe was installed across the dam base (see Fig. 3.63) to divert a local stream. Thereafter the pipe was plugged at the upstream end by concrete with the remaining length left open. The tailings dam was raised in stages above the Starter Dam. A supernatant pond covered tailings deposits upstream of the dam. Surplus water was reclaimed by a floating barge decant system and pumped back to the mill for reuse.

Cyanide-Laced Supernatant Water Spill on August 19–23, 1995

In the midnight of August 19, 1995 a stream of water was observed by a truck driver, issuing from the south end of the tailings dam. At dawn, the next day another discharge was observed at the north end of the dam together with extensive cracking along the dam crest. The combined discharges to the Omai River reached 50 m^3/s. The mine quickly diverted one stream into the open pit. The other stream, however,

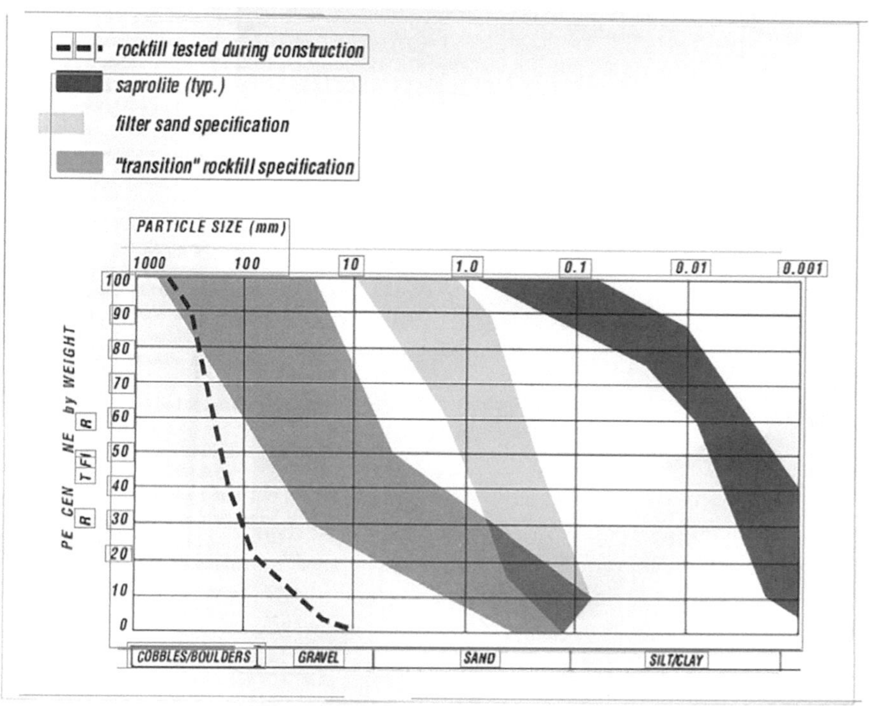

Fig. 3.57 Grain-size curves for damfill materials (from Vick, 1996)

required the construction of a cofferdam to divert it into the pit, taking about 100 h. In the end, 1.3 million m^3 of cyanide-laced supernatant water containing 25 ppm of total cyanide was captured in the pit. The remaining 2.9 million m^3 of water escaped into the Omai River and then the Essequibo. The spill resulted in environmental, ecological, socio-economic and political consequences for the affected community.

The Guayanese Government declared the affected region an environmental disaster area, and called for international assistance in addition to suspend mining operation at Omai. International governments and agencies that came to help Guayana to monitor river water quality, health, mining operation and dam safety issues included: Canadian and Venezuelan Governments, Pan American Health Organization and United Nation Department of Humanitarian Affairs (UN DHA, 1995).

Forensic Investigation on Causation of Omai Tailings Dam Failure

A dam review team consisted of members from Guayana, Argentina, UK, Canada and USA chaired by Vick (1995, 1996) was tasked to investigate the cause of the tailings dam failure. The team conducted site visits, interviewed technical staffs and reviewed dam design and construction records. They also carried out field investigations including trenching and drilling to supplement existing database as well as additional modelling and analyses to clarify and confirm their assessments. The following is a summary of their findings:

- The dam failed by internal erosion.
- The filter sand gradation was not compatible with the adjacent downstream rockfill (see Fig. 3.57). The filter zone was also too narrow (see Fig. 3.56).
- The upstream concrete plug of the decommissioned diversion corrugated-steel-pipe might have sections infilled with sand only. The pipe did not have seepage collars to increase seepage path along the contact of the pipe with the surrounding soil. Backfill around the pipe might not have carried out adequately to carry structural load of construction equipment and damfill.
- The workmanship, quality control and quality assurance of construction was not satisfactory.

The upstream embankment zones are shown on Figs. 3.58 and 3.59. Contact of filter/downstream rockfill zone is shown on Fig. 3.60. Core cracks and a sinkhole at the contact of upstream riprap/core zone revealed by trench excavation are respectively shown on Figs. 3.61 and 3.62. Figure 3.63 illustrates softened core zones and voids revealed from drill holes, and Fig. 3.64 shows disturbed dam fill revealed from test trench superimposed on dam-deformation pattern obtained from numerical analyses. The relative locations of the Omai mine site and two main rivers impacted by its supernatant water spill are shown on Fig. 3.65. Figure 3.66 illustrates results of river water quality monitoring program downstream of the spill.

Development After Spill

Due to the fact that the Omai mine gold production contributed to about 25% of the government revenue, there was enormous pressure to shorten the suspension period

Fig. 3.58 Tailings/riprap/core/filter/rockfill view along dam axis (from Vick, 1996)

Fig. 3.59 View of dam in opposite direction to that shown on Fig. 3.58 (from Poole, 2008)

Fig. 3.60 Close-up of filter/rockfill contact (from Vick, 1996)

Fig. 3.61 Core cracks exposed in trench excavation (from Vick, 1996)

Fig. 3.62 Close-up of exposed sinkhole in upstream riprap/core contact (from Vick, 1996)

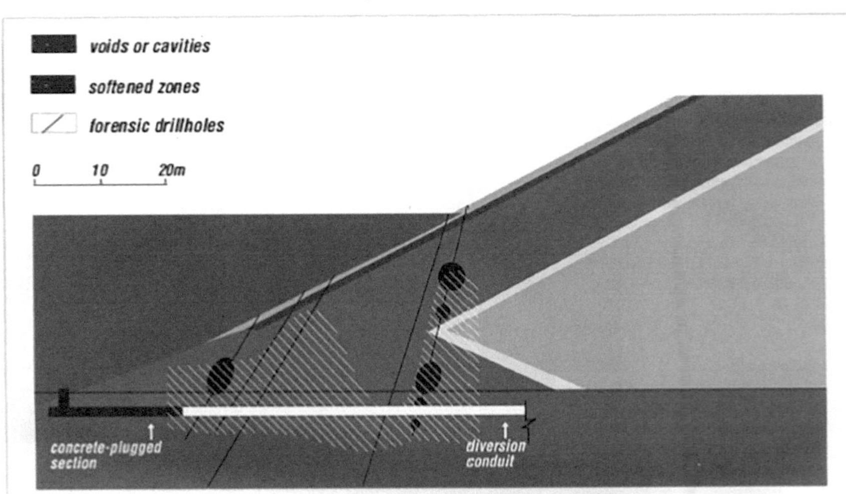

Fig. 3.63 Extent of softened core and voids revealed by inclined drill holes (from Vick, 1996)

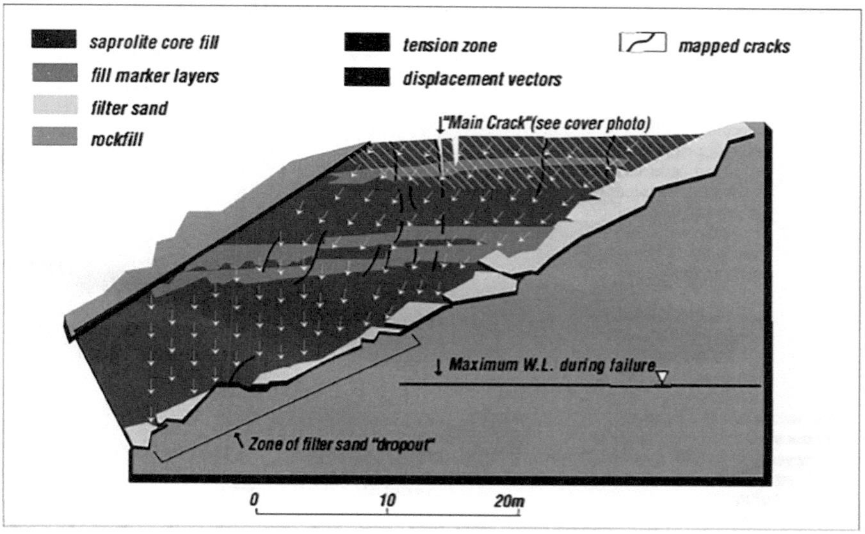

Fig. 3.64 Disturbance of damfills revealed in trench excavation (from Vick, 1996)

of its mining operation. The mine resumed its operation in March, 1996. Tailings storage shifted to a new impoundment, No. 2 Tailings Impoundment retained by six dams. A much robust No. 2 Tailings Dam has a cross section shown in Fig. 3.67. The dam has a wide central saprolite core, an upstream rockfill riprap, a crest and downstream rockfill protection zone, a central chimney and base blanket drain as well as a filter zone between the core and downstream rockfill zone. Special care was taken to ensure gradation compatibility at contact between adjacent zones. A filter geofabric was added between the filter sand and downstream rockfill as shown in Fig. 3.68. Figure 3.69 shows the vegetated No. 2 Tailings Impoundment after closure, while the tailings deposits in the failed No. 1 Tailings Impoundment had to be covered with rockfill to enhance stability after closure (Poole, 2008). In 2007 Iamgold acquired Cambior/Omai Gold Mine.

Lessons Learned

The 1995 Omai supernatant-water spill from the No. 1 Tailings Impoundment and the subsequent successful design, construction and closure of the No. 2 Tailings Impoundment highlighted following key lessons:

- The spill was caused by the loss of integrity of the impervious core of the No. 1 Tailings Dam due to migration of fine particles of saprolite clay core material into the downstream rockfill zone.
- The core failure was, in turn, resulting from a major design deficiency, i.e., use of relatively narrow filter zone of sand with a gradation incapable of stopping the migration of core material into the voids of the rockfill downstream of the filter.

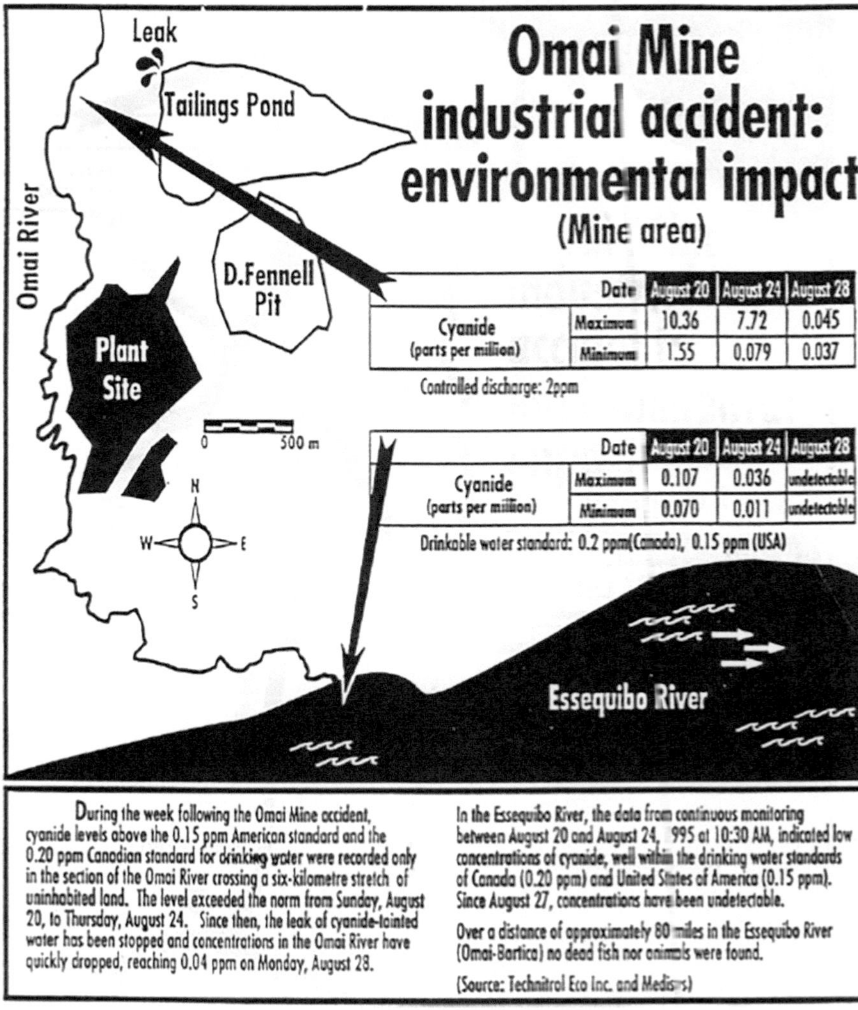

Fig. 3.65 Omai mine site and Omai River and Essequibo River impacted by its supernatant water spill (from Poole, 2008)

- The embedment of a 900 mm diameter corrugated steel pipe, used to divert a local stream during construction of the Starter Dam, within the dam base might have compounded the problem of internal erosion of the core material around the pipe.
- The mining operation including the design and construction of the No. 1 Tailings Impoundment appeared to be sub-standard.
- The contrast of the execution and performance of the No. 2 versus No. 1 Tailings Impoundment is remarkable. The operators are to be commended for their efforts and achievements in getting their acts together under the international spotlight in a short period of time.

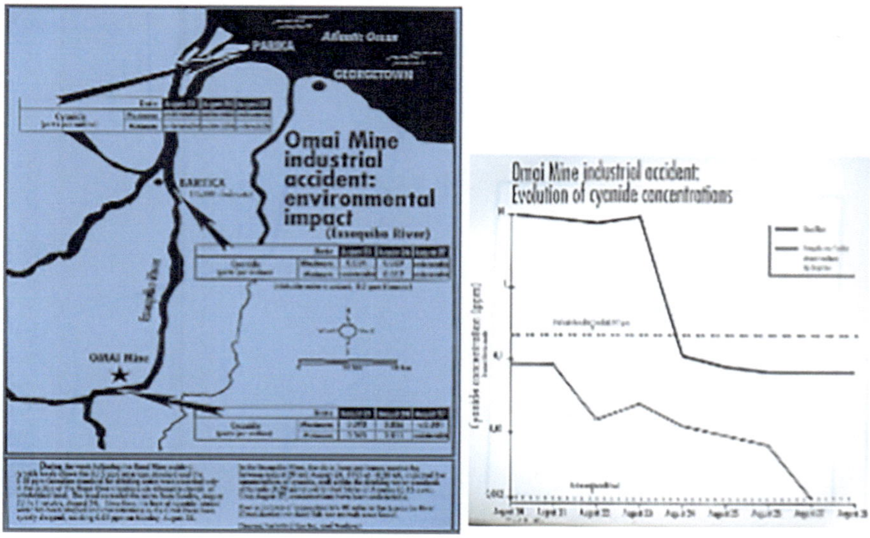

Fig. 3.66 Water quality monitoring along downstream rivers (from Poole, 2008)

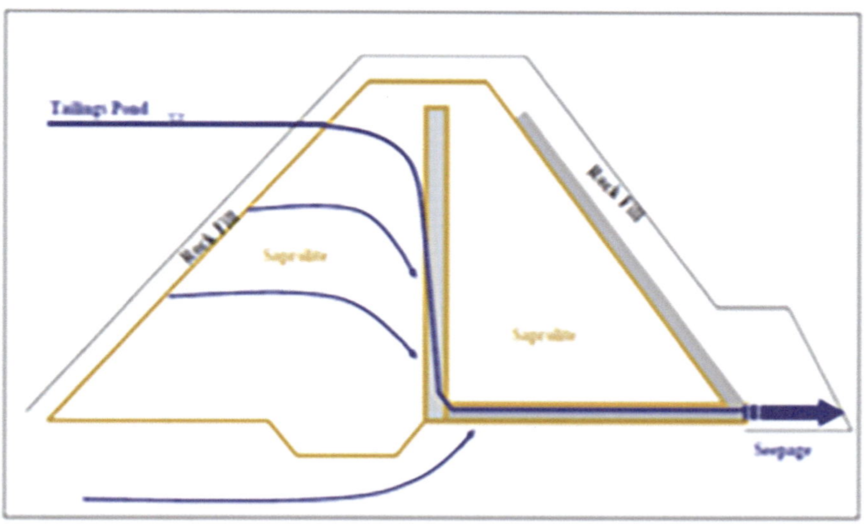

Fig. 3.67 Cross section of tailings Dam No. 2 constructed after spill incident in 1995 (from Poole, 2008)

Fig. 3.68 View of No. 2 tailings dam along dam axis (from Poole, 2008)

Fig. 3.69 No. 2 tailings impoundment after closure (from Poole, 2008)

3.11 Case 10 March 26, 1996 Marcopper Tailings Spill, Philippines

Marcopper Mine was a Canadian company that officially started its copper mining operations in 1969 at the Mt. Tapian Ore Deposit on Marinduque Island in the Philippines. When the Mt. Tapian reserve was depleted in 1990, Marcopper moved its operations to the San Antonio Ore Deposit, three kilometers north of the Mt. Tapian site (see Fig. 3.70). Mine tailings from the Mt. Tapian open pit operations were discharged into Calancan Bay located on the island's north shore. It is estimated that 84 million metric tonnes of mine tailings were discharged into the shallow bay. Due to the objection of this marine tailings disposal practice by local residents, the exhausted Mt. Tapian Pit was used to store tailings generated by the San Antonio operations after the drainage adits were sealed by concrete plug (see Fig. 3.71). A tailings spill incident occurred in 1996 due to failures of the sealing concrete plugs at the Tapian Pit lake drainage adits.

Fig. 3.70 Map of Tapian and San Antonio Ore Deposits, Marinduque Island, Philippines (from USGS, 2000)

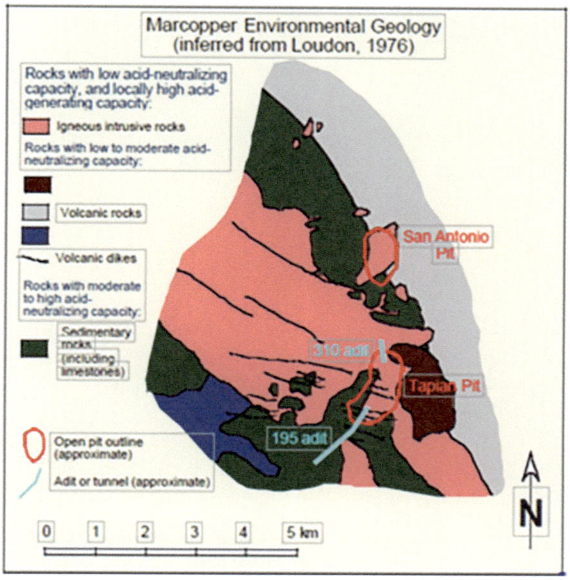

Fig. 3.71 Tapian Pit lake, 195 and 310 adits and Antonio Pit (from USGS, 2000)

Long-Term Environmental Pollution Caused by 1996 Boac River Tailings Spill

USGS (2000) sent a team to visit Marinduque Island on May 12–19, 2000 to get an overview of the 1996 Boac River tailings spill. From information gathered on their trip and the results of simple leach experiments they performed to examine metal mobility from the tailings, they concurred with conclusions of earlier studies by the mining company that the tailings deposits in the Boac River from the 1996 spill would be a long term source of acid and metals into the environment, and were in need of remediation.

Due to oxidation of sulfides in the tailings, the generation of acid waters during rainstorms, and subsequent evaporation of these acid waters during dry period, substantial deposits of soluble salts had built up within the tailings. These salts store acid and metals in a readily soluble solid form. During the next rainstorm, these salts were likely to dissolve and produce an ecologically damaging flush of acid and metals into the Boac River. The process of salt formation and dissolution could be repeated during each dry and wet cycle. Figures 3.72 and 3.73 show photos of tailings deposited in the middle reach of the Boac River, while Fig. 3.74 shows the medical examination of a resident's skin for possible signs of pollution impact. Recent satellite imagery of Tapian and Antonio Pit Lakes is shown in Fig. 3.75.

Lessons Learned

Tailings spilled to the Boac River forms the long-term source of acid and metal pollution. With each cycle of dry and wet period, the build up of soluble salt and its subsequent dissolution produces ecologically damaging flush of acid and metals into the Boac River. Thus, the 1996 tailings spill from the Tapian Pit Lake due to failed

Fig. 3.72 Tailings deposit in the middle Boac River (from USGS, 2000)

Fig. 3.73 Closeup of deposited tailings (from USGS, 2000)

Fig. 3.74 A local resident was examined for potential presence of skin lesions (from USGS, 2000)

Fig. 3.75 Recent satellite imagery of Tapian and Antonio Pit lakes (from Google Earth Pro)

sealing concrete plug at drainage adits created the long-term source of periodical flushes of acid and metals along the Boac River.

3.12 Case 11 April 25, 1998 Aznalcóllar Tailings Impoundment—Dam Foundation Failure, Spain

Los Frailes mine is located about 35 km west of Seville and 50 km north of Doñana National Park within the Iberian Pyrite Belt near the town of Aznalcóllar in the southwest of Spain. In the morning of April 25th, 1998 its tailings dam failed, and caused a spill of about 1.3 million m^3 of tailings and 5.5 million m^3 of supernatant water into the River Agrio and further down the River Guadiamar threatening to contaminate the Doñana National Park, an important migratory bird habitat and UNESCO world heritage site. Fortunately, there was no fatality. A total of 4600 ha of land were affected by the flood, and about 2600 ha were covered by tailings; 62% of these tailings were deposited in the first 13 km downstream of the dam. The flood event caused a massive fish kill and destroyed other aquatic life. The flood was prevented from entering into the Doñana Park by the construction of emergency containment dams (WWF, 2002).

The tailings impoundment, situated on the lower portion of a valley slope, is about 2 km long and 1 km wide elongated in the north–south direction. A L-shaped tailings dam along the east and south perimeters closes the impoundment with the natural contours forming the other two sides. The impoundment is subdivided by a divider dyke into two cells: the pyroclastic cell to the north and the pyrite cell to the south. The tailings dam was designed in 1978, and dam construction started in 1979 and reached the maximum height of 27 m at the time of failure. The dam was raised in stages over 20 years according to the operational requirement for tailings storage. In 1996, the original design dam height of the pyrite cell was increased by 1.39 m to the same crest elevation as the pyroclastic cell. Dam instrumentation consisted of inclinometers, piezometers and surface survey points.

The dam foundation subsoil profile consists of 3–5 m thick river terrace deposits overlying 70 m thick blue marl sediments (see Fig. 3.76). The blue marl, locally known as "Margas azules del Guadalquivir", is heavily over-consolidated, fissured, carbonate-rich clays of high plasticity. The marl is generally homogeneous on a large scale with no pre-existing failure planes. It contains systematic sub-vertical joints and a laminated texture associated with bedding planes. Random and discontinuous slickensides are present at variable frequency. The blue marl displays brittle behaviour when sheared rapidly under undrained condition, i.e., its shear strength drops substantially after sheared beyond the strain corresponding to the peak strength. Moreover, pore pressure in the marl foundation increased in response to the load increase of the damfill during dam raise and dissipated slowly with time. These characteristics of the blue marl foundation played a vital role in the eventual failure of the tailings dam.

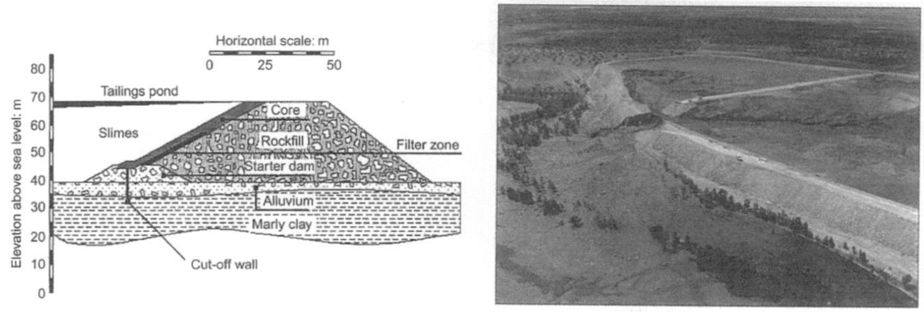

Fig. 3.76 **a** Section and **b** plan views of failed Aznalcóllar Dam (from Olalla & Cuellar, 2001)

On April 25, 1998 the dam breach occurred in the eastern tailings dam of the pyrite cell. The main supernatant pond was located at the southwest corner of the pyrite cell. The spill flowed through the breached divider dyke and the adjacent tailings dam of the pyroclastic cell at the southeast corner (see Fig. 3.76b), and eroded the tailings dam and its alluvial and blue marl foundation materials in the process. The dam failed as the result of a vary large horizontal displacement (reaching the maximum value of 60 m) of a 700 m long segment of the eastern tailings dam of the pyrite/ pyroclastic cell along a failure plane within the blue marl about 14 m below the dam/foundation interface. Upstream of the displaced dam block, the tailings surface dropped by about 4 m as the liquefied tailings flowed to infill the void evacuated by the downstream movement of the dam (see Fig. 3.77).

In the original 1978 design, the drained shear strength of the blue marl was assumed to be 25°. In the 1996 review, some minor changes of design parameters were made including the revised drained strength of the blue marl being 0.2 kg/ cm^2 and 22°. However, stability back analyses of the failed dam section using appropriate pore water pressure values indicated that the actual available strength of the blue marl at the time of incipient failure was probably substantially less than the adopted design strength. This overestimation of the available shear strength of the

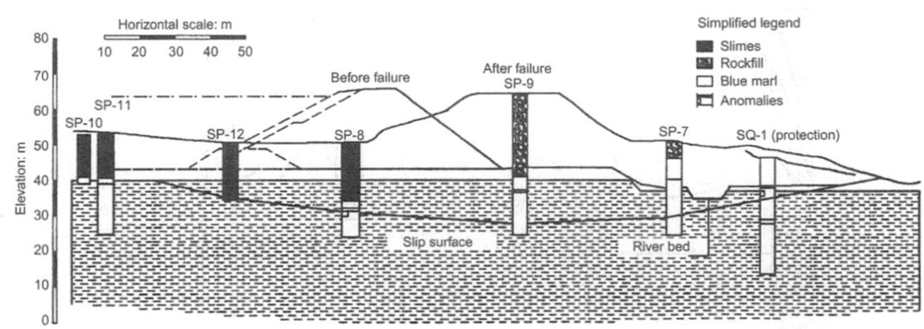

Fig. 3.77 Dam section profile before and after failure (from Olalla & Cuellar, 2001)

blue marl at the design stage was not rectified during the construction stage. Had the Observational Method been followed, using the up-to-date instrumentation data, the pre-failure foundation displacement would likely have been detected. However, missing this vital observation during the critical dam construction period sowed the seed of the eventual dam failure.

The dam breach released approximately 1.3 million cubic metres of fine pyrite tailings and 5.5 million cubic metres of tailings water into the Agrio and Guadiamar river systems. Table 3.7 shows chemical characteristics of estimated concentrations of dissolved metals in tailings water and elements in tailings, respectively (McDermott & Sibley, 2000)

A major clean-up program was mounted subsequent to the dam failure. Most of the escaped tailings were excavated and transported to the abandoned Aznalcóllar Open Pit for disposal. Contaminated water was retained, and treated before release. The total cost of various remedial measures reached 378 million Euros (WWF, 2002), which was borne by:

- Mining company—96 million Euros including clean-up of tailings close to the mine, indemnifying farmers for the loss of 1998 harvest, decommissioning the breached tailings impoundment and loss of mining opportunity;
- Regional government—145 million Euros, including clean-up of tailings in areas further away from the mine, purchase of polluted land and land restoration; and
- Central government—137 million Euros, including tailings removal, water treatment and river restoration.

Table 3.7 Estimated concentrations of dissolved metals in tailings water and elements in tailings (from McDermott & Sibley, 2000)

Dissolved metals in tailings water		Elements in tailings		
Element	Milligrams/litre	Element	Minimum %	Maximum %
Cu	17	Cu	0.18	0.22
Pb	3.5	Pb	0.8	1.0
Zn	450	Zn	0.9	1.1
Fe (ferrous)	85	Fe	40.5	42
Mn	100	S	44	46
Ca	350	As	0.6	0.7
Mg	150	Ni	0.0015	0.0020
Cd	1.3	Co	0.0050	0.0060
Ni	0.8	Cd	0.0030	0.0040
Co	0.8	Hg	0.0025	0.0035
Cr	0.07			
As	0.18			
Hg	< 0:001			
SO_4	5000			

Fig. 3.78 Recent satellite imagery of Seville and Donãna National Park (from Google Earth Pro)

Other comparable cost estimates on clean-up and remedial measures are also given in literature. The mining company finally closed the mining operation at Los Frailes in December 2001. Figure 3.78 shows a recent satellite imagery of Seville and Donãna National Park in the vicinity of the failed Aznalcóllar Tailings Dam, and Fig. 3.79 shows the location of the Aznalcóllar (Los Frailes) Tailings Dam in the Guadiamar river basin.

3.13 Case 12 October 11, 2000 Big Branch Slurry Impoundment Internal Erosion "Piping" Failure, U.S.A.

On October 11, 2000 a 40 ha fine coal waste impoundment of the Martin County Coal's preparation plant near Inez, Kentucky failed suddenly probably because of piping internal erosion (see Fig. 3.80a), and released 1.16 million m^3 of pond water and fine coal waste via adjacent underground mine workings (see Fig. 3.80b), and discharged 0.98 million m^3 of fluid through two portals into two watersheds: Wolf Creek and Coldwater Fork, flooding about 45 km of local streams, and contaminating 120 km of streams in Kentucky and West Virginia. Although no fatality occurred in the event, the slurry release killed aquatic life in the Tug Fork of the Big Sandy River and its tributaries, and shut down six water supply plants in the local areas for

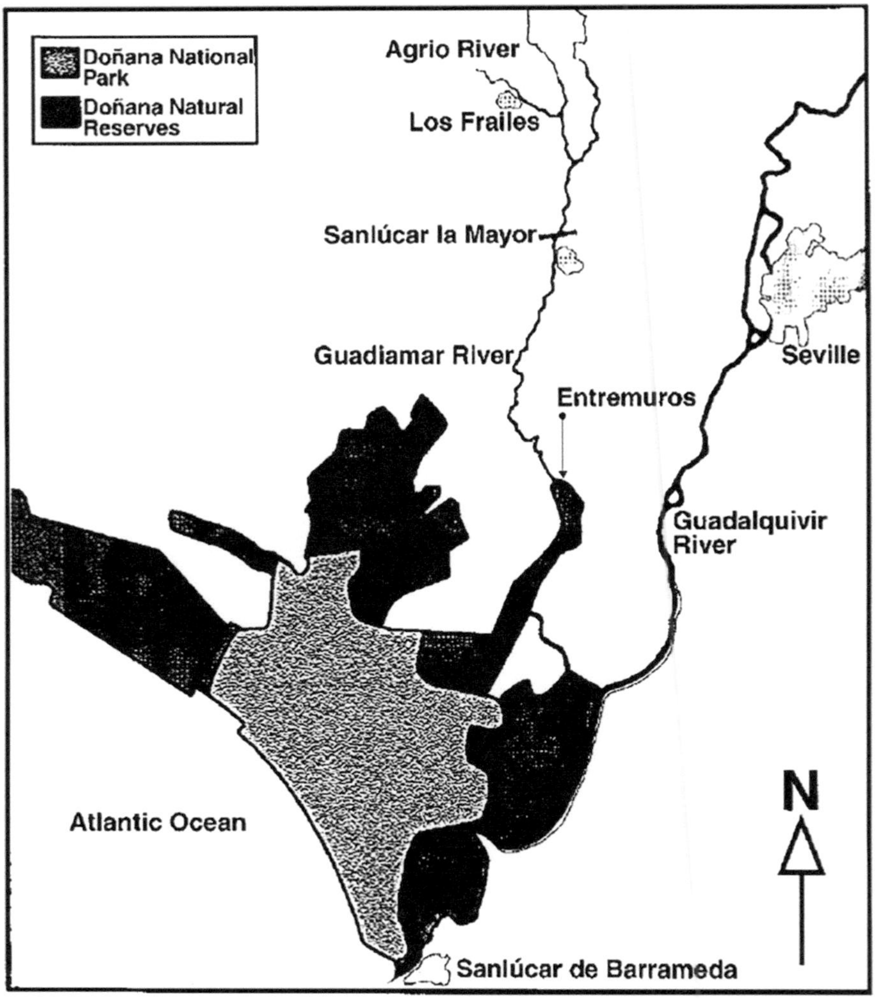

Fig. 3.79 Map of Guadiamar and Guadalquivir river systems (from McDermott & Sibley, 2000)

preventing contamination of potable water, costing more than $56 million dollars of damage. A similar event occurred earlier at the mine in May 1994 (Michael et al., 2000).

After the Big Branch impoundment incident, the U.S. Congress provided funding for the National Research Council (NRC) to examine ways to reduce the potential for similar accidents in the future. NRC (2002) published a book on "Coal Waste Impoundments—Risks, Responses, and Alternatives" to review the state of practice and prevailing regulatory framework, investigation, planning and mitigation of potential failures and alternative approaches for future coal waste disposal. Moreover,

Fig. 3.80 **a** Post-breakthrough overview of impoundment and **b** a portal breakout point (from Michael et al., 2000)

following the NRC recommendations, additional works resulted in the following two publications:

- "Guidance for Evaluating the Potential for Breakthroughs from Impoundments into Mine Workings and Breakthrough Prevention Measures" by a Work-Group of Federal and State Agencies (MSHA, 2003); and
- Update of "Engineering and Design Manual—Coal Refuse Disposal Facilities" (MSHA, 2009).

Lessons Learned

The generic risk of breakthrough of coal refuse slurry from its impoundment into underground mines was highlighted by Michael et al. (2000) and NRC (2002). Subsequently US Mine Safety and Health Administration issued guidance for assessing this risk and preventing measures (MSHA, 2003) and engineering and design manual (2009). Thus, this potential risk has been brought into proper regulative regime in USA.

3.14 Case 13 December 22, 2008 Kingston Fossil Plant Ash Dam Static Liquefaction Flow Failure, USA

Strictly speaking unlike coal waste, ash is the coal burning product rather than coal preparation by-product (tailings). However, ash storage shares many similar characteristics with coal fine waste (tailings) storage. For this reason, an ash storage facility failure is included in this section. Since ash contains trace concentration of rare earth elements, current research is also being conducted to seek economic means to extract these strategically important elements from ash waste (http://www.waste-management-world.com/articles/print/volume-14/issue-5/wmw-special-recycling-focue/coal-ash-recycling-a-rare-opportunity.html).

The Kingston Fossil Plant is a 1700 MW coal-fired power plant, located in Roane County, Tennessee on a peninsula formed by the confluence of the Emory River to the north and the Clinch River to the south and east. The plant, was reportedly as the largest coal-burning power plant in the world, when commissioned in 1955. It generates 10 billion kilowatt-hours of electricity a year, sufficient to supply the need of about 670,000 homes in the Tennessee Valley. The fly ash and bottom ash generated by the power plant is stored in a facility located over a former flood plain area near Harriman adjacent to the Watts Bar Reservoir, north of the plant (see Fig. 3.81 for the plant and ash storage areas).

Tennessee Valley Authority (TVA) is a public power company. It owns and operates Kingston Fossil Plant. After the December 22, 2008 failure of its ash dam, it engaged AECOM (2009) to perform a root cause analysis for the failure. In response to the root cause report, TVA (2009) issued its review on July 23, 2009. The following section is based mainly on these two reports.

Kingston Ash Storage Facility

A general layout of the Ash Storage Facility is shown in Fig. 3.82, and the north-south section of Dredge Cell 2 is shown in Fig. 3.83. The ash pond northern containment Starter Dyke C was built of nearby clay residuum and bottom ash by 1959. Dyke C was later raised by the upstream method with a side slope of about 4 H:1 V to an ultimate height of 35 ft. From 1954 to 1958 ash, river silt and Dyke C clay runoff was

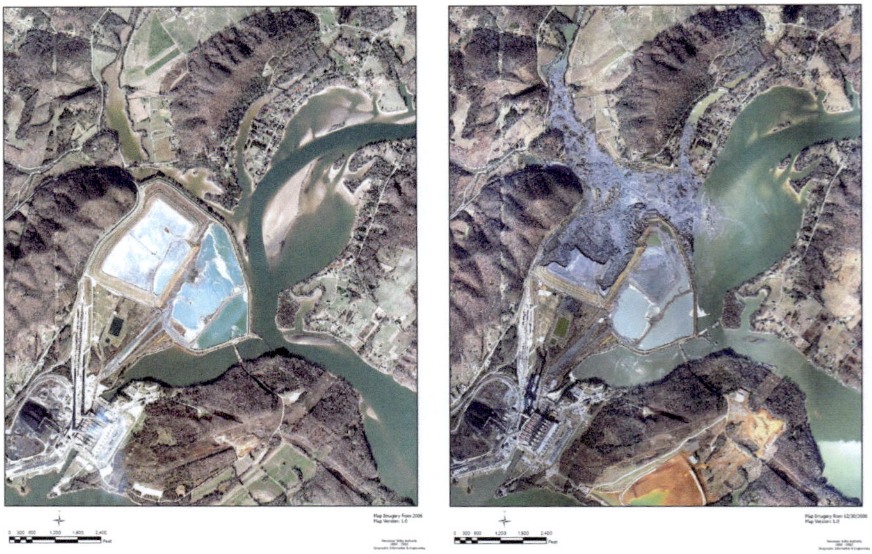

(a) Prior to Ash Dyke failure (b) After Ash Dyke failure

Fig. 3.81 Aerial view of plant and ash pond areas (from AECOM, 2009). **a** Prior to Ash Dyke failure. **b** After Ash Dyke failure

deposited over the slack water embayment under deltaic conditions over the northern half of the foundation area of the dredge cell that failed in 2008. Since 1958, mostly fly ash was first settled out in a stilling pond, and then dredged out to deposit within dredge cells (reaching about 85 ft. thick in 2008), which are contained by a series of dykes, with an average slope of 5 H:1 V, raised by the upstream method (see Fig. 3.84). During its operation up to 2008, the storage facility had gone through several incidents involving minor slope instability and seepage issues but no major failure.

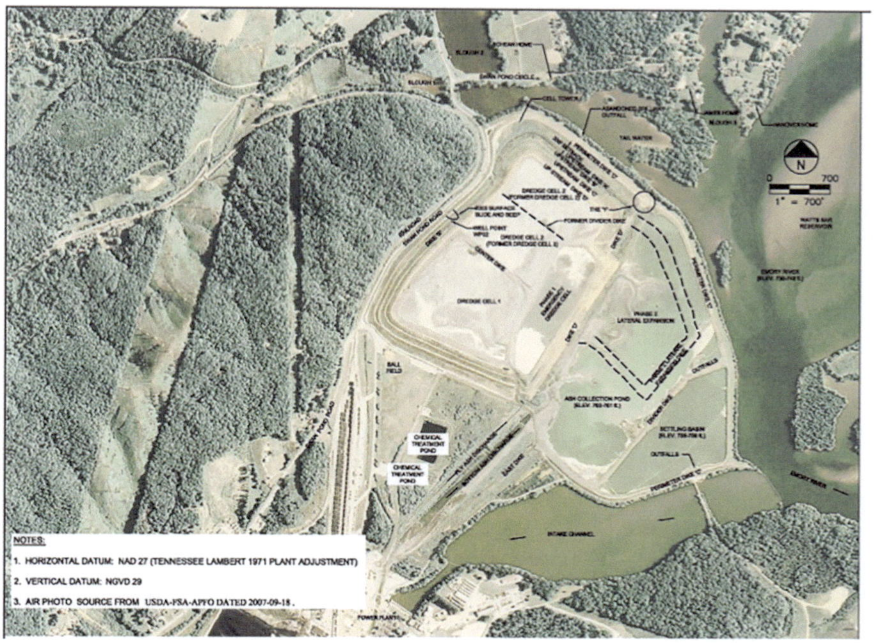

Fig. 3.82 Layout of ash pond dredge cells (from AECOM, 2009)

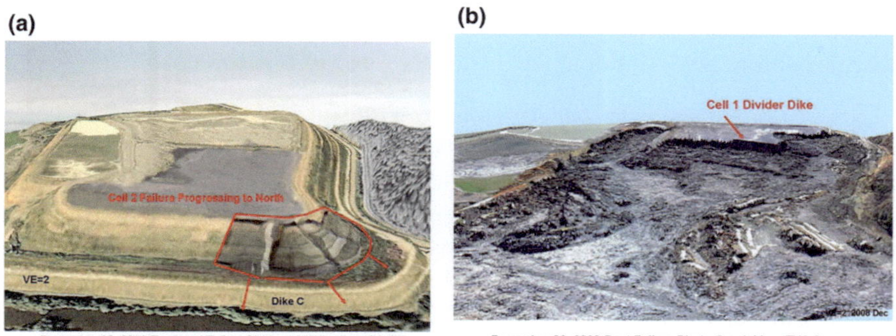

Fig. 3.83 **a** Initiation of Dyke failure and **b** post-failure Dyke condition (from AECOM, 2009)

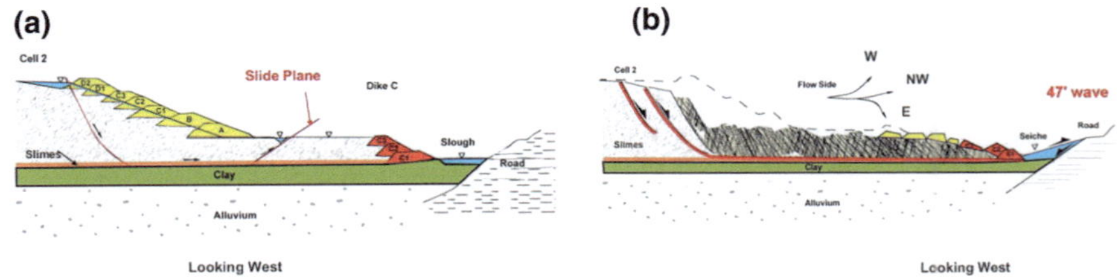

Fig. 3.84 **a** Initial failure of Dredge Cell 2 and **b** retrogressive failure of Cell 2 (from AECOM, 2009)

Failure Investigation

Between November 20 and December 21, 2008 there were no more than 8 in. of rain in the area. About 1 in. of rain fell in the day prior to failure, but no adverse conditions such as evidence of seepage, or shallow instability were observed during inspections at noon on December 21, 2008. On December 22, 2008 around 1:00 a.m. the north and central portions of Cell 2 suddenly failed. An estimated 5.4 million cubic yards of material, consisting primarily of hydraulic-filled ash and its peripheral containment dykes, were released in a sequence of retrogressive flow slides for about one hour. The flow slide ran out northward for about 3200 ft. beyond the original limit of Dyke C into the former Emory River Channel of Watts Bar Reservoir. Along its path, the slide mass inundated several TVA-owned sloughs, spread onto eight acres of private property and damaged several structures (see Fig. 3.81b). At the time of failure, there were no earthquakes recorded by local seismic stations. Fortunately, the land surrounding the power plant and ash pond is undeveloped and sparsely populated. The flow slide created a flood wave that ran both upstream and downstream of the surrounding waterways. It is estimated that 42 residential properties were affected, and 22 residences were evacuated. However, there were no fatalities or injuries.

The dredge cells foundation subsoil profile consists of:

- about 6 in. thick layer of thin laminas (a few mm thick) of dark grey fine fly ash and reddish brown clayey silt;
- soft to stiff clays;
- alluvial sand and silts; and
- shale bedrock.

The top layer of laminated material, referred to as slimes, is the weakest foundation material. The slimes are sensitive silt with Liquidity Index in the range of 4.7–7. Its Liquid Limit ranges from 30 to 60, and Plasticity Index ranges from 10 to 25. The undrained shear strength of slimes develops a peak at 5% shear strain, thereafter it gradually reduces to 75% of the peak strength at 20% strain value. Moreover, the slimes are subject to creep under sustained shear stress equal to or greater than 80% of its peak shear strength.

As shown in Fig. 3.84, the initial failure plane passes through the contact between the stored ash and foundation slimes. The ash material contains non-plastic silt size particles of spherical shape. The void ratio of unfailed ash is about 0.84–0.88. Its drained shear strength has a friction angle of 30°. Under undrained triaxial shear test, it develops a peak strength at 0.5% strain, thereafter the strength rapidly decrease to the steady-state value of 100 psf or less. Thus, when the ash pond is being raised gradually, the ash material consolidates and deforms (shears) close to a drained condition. However, under more rapid loading condition the ongoing creep of the foundation slimes could increase shear stress on the ash material under undrained condition. The sudden drop in the shear strength of the ash from the drained value to undrained value would trigger a flow-slide failure, and this is generally recognized as a static-liquefaction phenomenon.

The post-failure investigation by AECOM (2009) indicated that the failure most likely initiated in the northwest corner of Dredge Cell 2 where perimeter Dyke C retained a 200 ft. wide low-level ash berm. Undisturbed samples retrieved from boreholes in the area show that the foundation slimes were the seat of slide movement. It is inferred from stability analyses that the initial failure plane daylighted through the low-level ash berm (see Figs. 3.83a and 3.84a). This initial failure triggered rapid retrogressive slope failures of Dredge Cell 2, and breached perimeter Dyke C (see Figs. 3.83b and 3.84b).

TVA (2009) Review of AECOM (2009) Root Cause Report

The TVA (2009) document consists of:

Executive Highlights
Introduction
Summary of Findings
Factual Background
Findings
Recommendations
Appendices
Appendix A Objective, Scope and Methodology
Appendix B Marshall Miller and Associates Report on Root Cause Analysis
Appendix C Other Observations by Marshall Miller
Appendix D Scope and Methodology as Defined in the Root Cause Analysis Report
Appendix E Memorandum Dated July 20, 2009, from Tom Kilgore to Richard W. Moore

Essentially, the TVA document covers the peer review of AECOM (2009) report by Marshall Miller as well as the management aspect of TVA Ash Ponds in general. A third consultant, Stantec, was also engaged by TVA to assist in addressing the overall safety issue of Ash Ponds.

Post-failure Development

Ash deposited in the river caused elevated arsenic and selenium levels in water. Thus, the first phase of clean up was dredging ash out of river channels, and shipping the material to a landfill in Perry County, Alabama. The second phase was to remove ash deposited on land. The clean-up of Kingston Ash Pond failure was still ongoing in December 2013 (http://www.usatoday.com/story/news/nation/2013/12/22/coal-ash-spill/4143995/), and its cost was estimated to be in the order of $1.2 billion, when the work is completed in 2015.

A little more than 5 years after the Kingston Ash Pond failure, on February 2, 2014 a break in a storm-water pipe beneath an ash basin at the retired Dan River Steam Station in Eden, North Carolina caused another release of about 22 million gallons of water and 30,000–39,000 tonnes of ash into the Dan River. The ash dam along the river was not affected, although some erosion had occurred inside the ash basin farthest from the river (http://www.duke-energy.com/news/releases/2014020301e.asp).

Since the Kingston spill, the Environmental Protection Agency (EPA) raised concerns about current ash storage practices in the United States, and initiated a general review of other ash storage facilities. Critical facilities were identified, and measures for improving these facilities were either implemented or are being addressed. Apparently, many of these facilities have been considered as landfills rather than tailings dams, and, therefore, are deficient in the dam safety aspect. The EPA has been working to strengthen the regulations on the design, construction and operation of these facilities. However, the process has been influenced by many stakeholders and involved both the Federal and State jurisdictions. Similar to the enactment of SMCRA (1977) as the reaction to the Buffalo Creek Coal Waste Impoundment Failure in 1972, more stringent federal regulations of ash impoundments are expected after the transition phase of ongoing consensus building of all stakeholders. Subsequent to the February 2, 2014 Eden ash spill in North Carolina, on September 20, 2014 Senate Bill 729 (the Coal Ash Management Act) became law without the Governor's signature aiming to improve the coal ash management practice in the state (http://www.smithenvironment.com/n-c-coal-ash-bill-becomes-law/).

On December 19, 2014, the EPA Administrator signed a final rule that establishes a comprehensive set of requirements for both existing and new Coal Combustion Residuals (CCRs or coal ash) in landfills and surface impoundments. These requirements have been finalized under the solid waste provisions, subtitle D, of the Resource Conservation and Recovery Act. The rule includes following requirements:

- Reducing Risk of Catastrophic Failure;
- Protecting Groundwater;
- Establishing Operating Criteria;
- Record Keeping, Notification and Internet Posting;
- Coordinating with State Programs; and
- Covering Closure and Beneficial Use.

Thus, almost exactly six years after the failure of Kingston Fossil Plant Ash Dyke, the EPA finally closes the federal regulation gap regarding the Disposal of Coal Combustion Residuals from Electric Utilities (EPA, 2014)

Lessons Learned

The sudden failure of the Kingston ash impoundments is due to static liquefaction of the perimeter ash retention dyke. The static liquefaction failure mechanism will become a familiar one to the readers of the Main Chapter on representative case histories.

The Kingston Ash Impoundment failure is an example of a generic risk posed by the storage of coal combustion residues (or coal ash) operated by fossil power plant. Although the risk is prevalent, it seemed to have not been consistently regulated by US federal agencies. Six years after the failure, the regulation gap was finally closed by the US Environmental Protection Agency (EPA, 2014).

3.15 Case 14 27 February 2010 Las Palmas Tailings Impoundment—Dam Seismic Liquefaction Flow Failure, Chile

Las Palmas Dam Construction

The Las Palmas Tailings Dams were built upon a sloping ground that sloping downward toward the south and east as shown in Fig. 3.85. The dams, located in the city of Pencahue, were constructed from 1981 to 1998 in four stages as shown in Fig. 3.86. At each stage, the Starter Dam and Crest-raising Dam were constructed using sand-fraction of tailings, while storage of tailings had been carried out by hydraulic fill method.

Fig. 3.85 Location of Las Palmas tailings dam shown on Google Earth satellite imagery and Google map

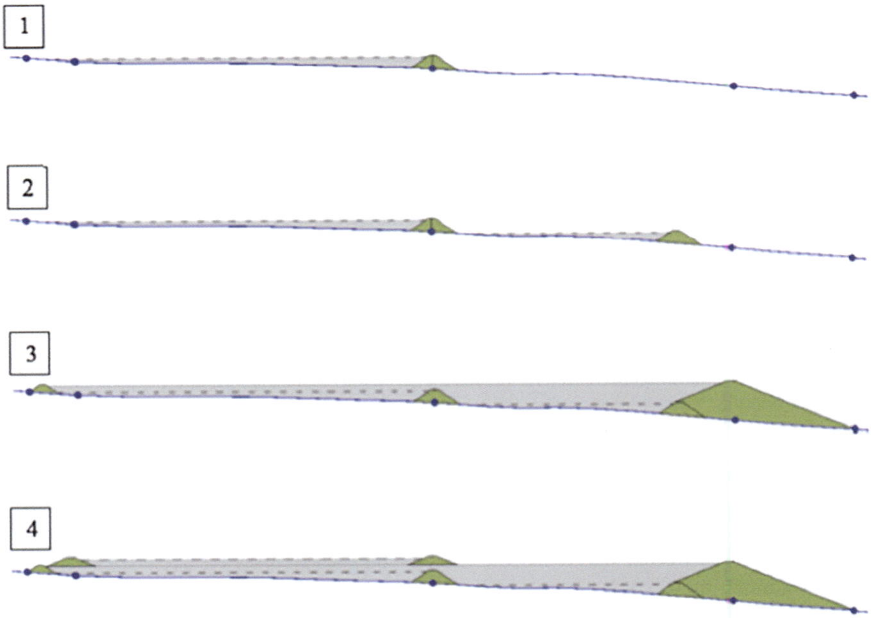

Fig. 3.86 Tailings dam construction in four stages from 1981 to 1998 (from PEER, 2019)

As shown in Fig. 3.86, the tailings impoundment had been constructed as outlined below:

- Stage 1 (1981–1986)—The upper impoundment was retained by a middle starter dam;
- Stage 2 (1986–1992)—The lower impoundment was retained by a downstream starter dam;
- Stage 3 (1992–1997)—The entire impoundment was retained by an upstream starter dam and a raised downstream dam. The middle starter dam appeared to have been inundated by deposited tailings;
- Stage 4 (1997–1998)—The upper impoundment was retained by the raised upstream starter dam and a newly constructed middle dam founded on deposited tailings. Moreover, the upstream starter dam appeared to be raised by the upstream construction method as interpreted from Fig. 3.86.

The tailings impoundment was closed in 1998 with a clayey soil cover. After closure, the tailings area was partially covered with a 150 mm thick layer of gravelly material.

Flow Failure of Liquefied Tailings Dam

The 27 February 2010 Maule, Chile, earthquake is the seismic event that resulted in the flow failure of liquefied downstream Las Palmas Tailings Dam. The strong ground motion recorded at the town of Talca is shown in Table 3.8. As the tailings impoundment is located about 20 km north-northwest of Talca (see Fig. 3.85), the ground shaking received at the damsite could be represented by the earthquake parameters shown in Table 3.8.

The strong ground shaking resulted in liquefaction of susceptible tailings material. Figure 3.87 shows the tailings impoundment before and after the dam flow failure. Figure 3.88 shows an aerial photo of the failed impoundment, while Fig. 3.89 shows the areal extent of the intact tailings retained in the upper impoundment and the failed tailings flowed from the lower impoundment. Villavicencio et al. (2014) discussed Las Palmas tailings impoundment failure in their overview of Chilean sand tailings dams. They indicated that the downstream tailings dam failed due to seismic liquefaction of dam material at the base level. There was a saturated zone of 0.5–1 m due to incoming groundwater undetected at the time of closure. Figure 3.90 shows the collapsed tailings impoundment, and Fig. 3.91 shows groundwater seeping from a hill slope adjacent to the Las Palmas tailings dam. Villavicencio et al. (2014) also commented on Chilean guidelines on tailings management practice. In 2007, Supreme Decree No. 248 was promulgated to replace Supreme Decree No. 86 promulgated in 1970. It improved many aspects of the older decree.

Figure 3.92 shows tailings sand-boil along a flow path. The leading front of the easterly flow traveled approximately 165 m (540 ft.), whereas the southerly flow front traveled roughly 350 m (1150 ft.). The total volume of displaced material was estimated about 231,660 m^3 (303,000 yds^3) (Gebhart, 2016).

Gebhart (2016) used Slope/W software to model the progressive failure of the downstream tailings dam. Figure 3.93 illustrates the sequence of failures of the tailings impoundment. The failure of the middle dam is quite understandable, as it is founded on deposited tailings. The fact that the upstream dam survived the earthquake may be due to its location (on the upstream end of the impoundment with the original ground surface sloping towards downstream, hence retaining tailings of least thickness) and foundation condition (foundation could be relatively more free-draining). These assumed favarouble conditions potentially could result in thinner saturated tailings deposits against the upstream dam. Furthermore, with lower height, both the upstream dam and the retained tailings are less contractive than their counterparts located in other areas of the impoundment.

Lessons Learned

There are limited public documents available for this case. The following lessons are gleaned from four documents listed in the reference section:

- Villavicencio et al. (2014) offered a plausible rationale for why the downstream dam failed by seismic liquefaction. Low density of damfill due to lack of adequate compaction and its saturation near the dam base by incoming groundwater flow could explain why the dam failed by seismic liquefaction.
- The post-earthquake studies provided technical information and back analyses of the dam to shed some lights about the liquefaction flow slide of the lower tailings impoundment. It also documented the post-liquefaction residual undrained shear

Table 3.8 Ground motion recorded at Talca (after Boroschek et al., 2012)

Code station (Rrup)	Channel	PGA (g)	PGV (cm/s)	PGD (cm)	5% damped PSa (g)					Arias intensity (m/s)	Significant duration (s)	Central frequency (Hz)
					0.1 s	0.2 s	1 s	2 s	3 s			
TAL (66 km)	L	0.48	28	4	0.77	1.22	0.31	0.13	0.05	11.61	69.9	5.14
	T	0.42	34	7	1.01	1.79	0.38	0.19	0.08	11.06	71.9	5.14

(a) Bird's-eye view of Tailings Impoundment Prior to Failure (b) Upper Scarp of Failed Impoundment

Fig. 3.87 Las Palmas tailings impoundment prior to and after seismic liquefaction flow failure (from GEER, 2010). **a** Bird's-eye view of tailings impoundment prior to failure. **b** Upper scarp of failed impoundment

Fig. 3.88 Aerial photo of failed tailings impoundment (from PEER, 2019)

strength of the failed tailings being in the range of 7.8 kPa (163 psf) to 8.7 kPa (181 psf).

3.16 Case 15 August 4, 2014 Mount Polley Tailings Impoundment Dam Foundation Failure, Canada

Mount Polley Mine is located in south-central British Columbia, 8 km southwest of Likely and 56 km northeast of Williams Lake. The mine is an open-pit operation mining an alkalic copper–gold porphyry deposit with a capacity to process 18,000 tonnes of ore per day.

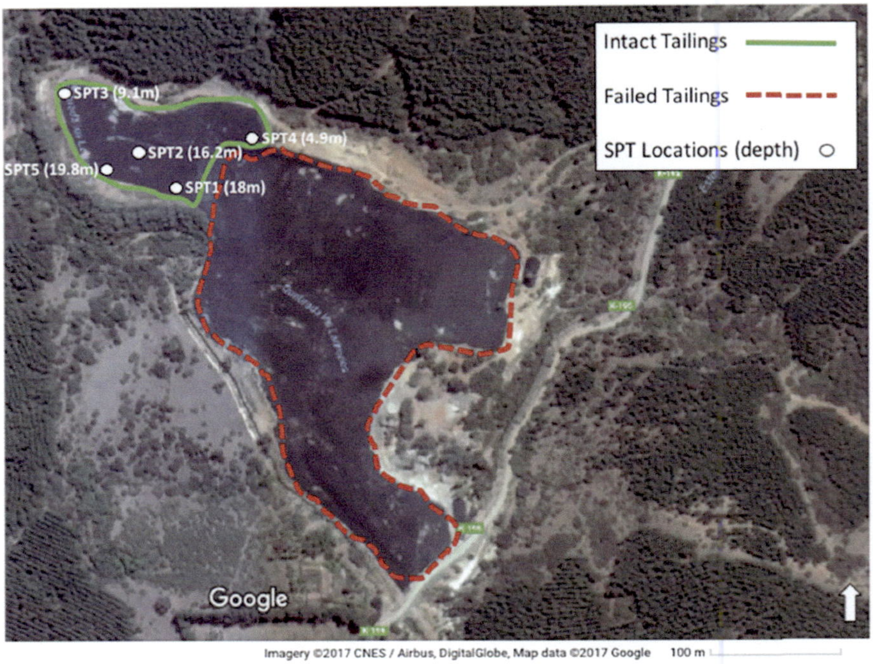

Fig. 3.89 Areal extent of intact and failed tailings (from PEER, 2019)

Fig. 3.90 Collapsed Las Palmas tailings impoundment (from Villavicencio et al., 2014)

Fig. 3.91 Groundwater flow through failed tailings dam (from Villavicencio et al., 2014)

In the morning of August 4, 2014 the tailings dam that retains the tailings impoundment breached (ENR, 2014) with a 100 m long gap (see Fig. 3.94), and released about 17 million cubic metres of water and 8 million cubic metres of alkaline tailings and earth materials (see Figs. 3.95 and 3.96). Fortunately, there was no loss of life, or personal injury.

Emergency Response

Immediately after the dam breach, the Cariboo Regional District issued the water advisory to warn the public to not drink water in the affected areas, which covered the entire Quesnel and Cariboo Rivers system, right to the Fraser River. By August 12, the water advisory had been lifted with the exception of the locally impacted area including Polley Lake, Hazeltine Creek and its mouth at Quesnel Lake.

The downstream environmental impact of the breach includes:

- erosion and scour along Hazeltine Creek;
- deposition of trees and woody debris in Polley Lake, along Hazeltine Creek erosion path and Quesnel Lake near the mouth of Hazeltine Creek; and
- deposition of tailings and eroded earth material in Polley Lake, Hazeltine Creek, and Quesnel Lake.

Fig. 3.92 Tailings sand-boil along flow path (from GEER, 2010)

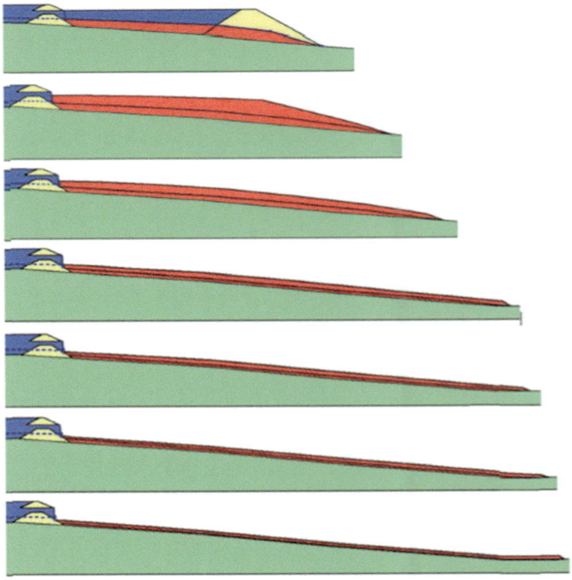

Fig. 3.93 Modelled progressive failure of lower tailings impoundment (from Gebhart, 2016)

(a) Upstream tailings inside breached dam

(b) Looking from downstream of dam breach

Fig. 3.94 Aerial views of breached Mount Polley tailings impoundment (http://www.vancouversun.com/news/Photos+Mount+Polley+Mine+tailings+break/10094276/story.html). **a** Upstream tailings inside breached dam. **b** Looking from downstream of dam breach

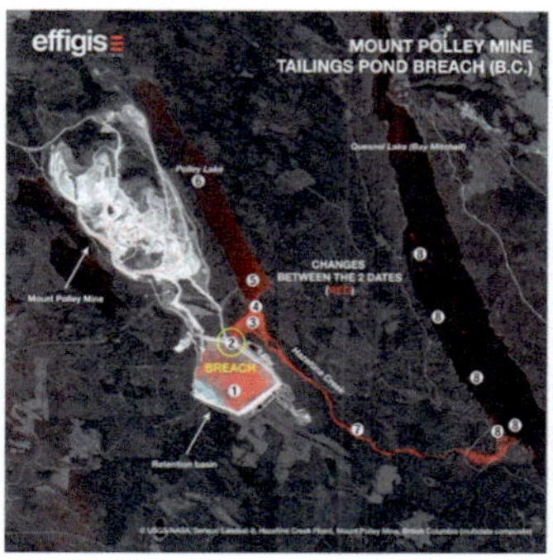

LEGEND:

1. Retention basin is emptied of its water and part of its tailings
2. Breach in the retention basin dike
3. Section of forest destroyed by the flow of water and mud
4. Wetlands covered in mud
5. Part of Polley Lake containing a high concentration of suspended matter
6. Part of Polley Lake containing an average concentration of suspended matter
7. Flow of mud has widened the bed of the Hazeltine Creek to one hundred metres or more
8. Plant debris (trunks) carried away by the flow of water and mud

Changes between July 29, 2014 and August 5, 2014

Fig. 3.95 Environmental impact of dam breach shown on composite satellite imagery (http://effigis.com/mount-polley-mine)

(a) Tailings deposited in upstream reach of Haseltine Creek and Polley Lake

(b) Tailings deposited in downstream reach of Haseltine Creek at mouth of Quesnel Lake

Fig. 3.96 Environmental impact on downstream recreational fish-bearing stream and lakes (http://www.vancouversun.com/news/Photos+Mount+Polley+Mine+tailings+break/10094276/story.html). **a** Tailings deposited in upstream reach of Haseltine Creek and Polley Lake. **b** Tailings deposited in downstream reach of Haseltine Creek at mouth of Quesnel Lake

On August 5, 2014, B.C. Ministry of Environment issued a *Pollution Abatement Order* (http://www.env.gov.bc.ca/eemp/incidents/2014/pdf/PollutionAbatementOrder_20140805.pdf) to the mine (http://www.env.gov.bc.ca/eemp/incidents/2014/mount-polley/). Since then it had overseen implementation of the mine's action plan in order to assess potential environmental impacts and to monitor ecosystem recovery.

The mine followed a Conceptual Interim Erosion and Sediment Control Plan, submitted to the Ministry, for mitigating ongoing erosion and sediment transport within downstream impacted areas with the following objectives (http://www.imperialmetals.com/s/Mt_Polley_Update.asp?ReportID=671041):

- provide water management structures to improve the quality of water flowing into Quesnel Lake;
- reduce the potential for re-mobilization of tailings and sediments that were deposited or exposed by the breach; and
- minimize and control flows from the tailings impoundment and re-direct these flows to the depleted Springer Pit.

The plan focused on the following three high-priority areas (see Figs. 3.95 and 3.96):

- Within and down-gradient of the tailings impoundment—to stop flow of water and tailings downstream by constructing Upstream and Satellite Dykes and water diversion facilities (see Fig. 3.97), and to remove wooden debris from Hazeltine Creek outlet and Quesnel Lake;

Fig. 3.97 Aerial view of emergency repair to contain upstream tailings after dam breach. *Image credit* B.C. Ministry of Environment, September 22, 2014

- Polley Lake—to lower lake level in order to prevent uncontrolled release of lake water by transferring excess water to Hazeltine Creek; and
- Upstream of the mouth of Hazeltine Creek to Quesnel Lake—to construct a series of settling ponds for removing suspended solids prior to draining water into the lake.

The mine also conducted various environmental sampling and testing programs to track the changes of water, sediments and their impacts on fish in the impacted areas.

On August 18, the Ministry of Mines and Energy, with the support of the Soda Creek Indian Band and Williams Lake Indian Band, ordered an *independent engineering investigation* (http://www.newsroom.gov.bc.ca/2014/08/independent-expert-engineering-review-launched-following-mount-polley-dam-breach.html) and inquiry into the Mount Polley tailings pond breach (https://www.mountpolleyreviewpanel.ca/). Moreover, it ordered independent third-party review of the 2014 Dam Safety Inspections at all 98 tailings impoundments in the province, including 31 active and 67 closed or under care and maintenance facilities.

Independent Review Panel Investigation

On January 30, 2015, the Mount Polley Independent Expert Engineering Investigation and Review Panel, consisting of internationally renowned Prof. Norbert Morgenstern—Chair, Mr. Steven G. Vick and Prof. Dirk Van Zyl, issued its comprehensive report (https://www.mountpolleyreviewpanel.ca/final-report). The following highlights the salient features of the panel's findings with the aid of Figs. 3.98, 3.99, 3.100 and 3.101 extracted from the report.

Figure 3.98a shows the plan view of the Mount Polly Tailings Dam that is comprised of three segments: Perimeter Embankment, Main Embankment and South Embankment with station numbers, Station 0 + 452 to 4 + 895, and dam section designations, Sections A to G, I to K as shown in the figure. Figure 3.98b is the enlarged plan view of the Perimeter Embankment in the vicinity of the breached segment from Station 4 + 200 to 4 + 300. Superimposed on the as-built embankment-fill surface black contours are the coloured contours showing the thickness in metres of the Upper Glacio-Lacustrine foundation unit, a relatively weak foundation layer at the depth of about 8 m or deeper below the original ground surface. Figure 3.98c, d show the respective Dam Section A with a downstream buttress berm in the Main Embankment and Dam Section D with no buttress berm in the Perimeter Embankment near Station 4 + 000, about 200 m southeast of the breached segment.

Photos of soil samples retrieved from the critical Upper Glacio-Lacustrine foundation unit are shown in Fig. 3.99. Figure 3.99a for sample from inside the dam breach zone, while Fig. 3.99b for sample from outside the breach. The lacustrine deposits distorted by dam movement in the breach zone shown in Fig. 3.99a contrasts strongly with the orderly laminated varves shown in Fig. 3.99b.

Figure 3.100a shows the back analysis of the downstream slope stability for the failed dam section in the breach zone using the undrained shear strength ratio of s_{un}

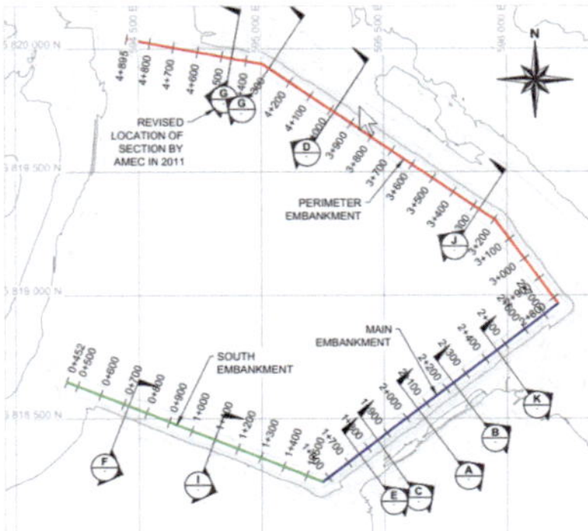

(a) Plan view of tailings dam

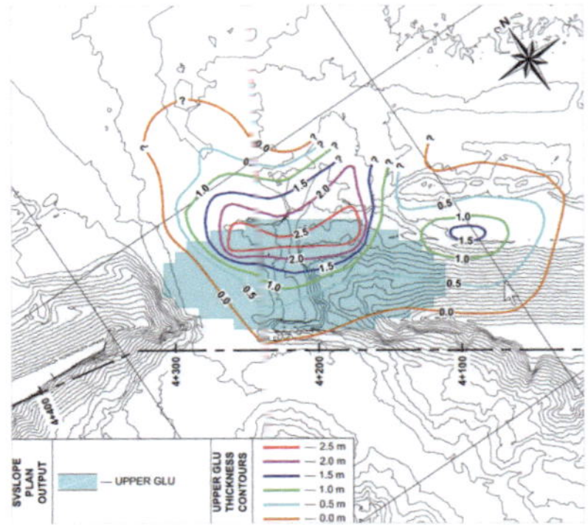

(b) Thickness contours of Upper Glacio-Lacustrine foundation unit in vicinity of breached dam segment

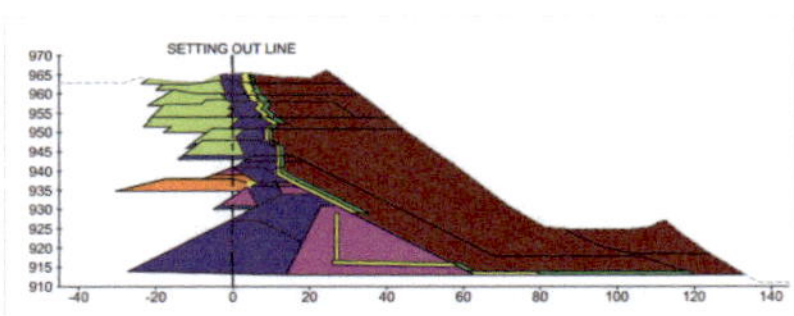

(c) Dam Section A in Main Embankment

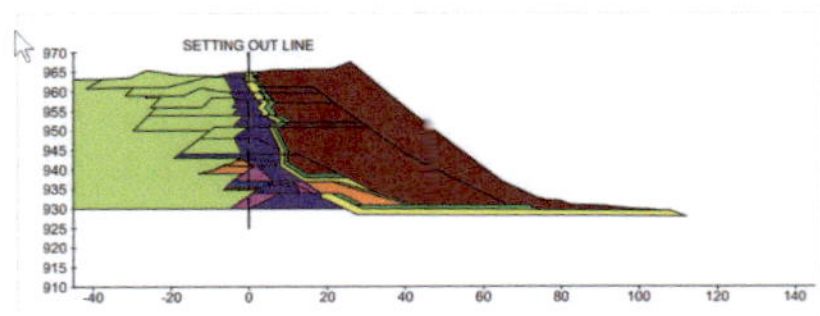

(d) Dam Section D in Perimeter Embankment

Fig. 3.98 Plans and sections of tailings dam (from https://www.mountpolleyreviewpanel.ca/final-report). **a** Plan view of tailings dam. **b** Thickness contours of Upper Glacio-Lacustrine foundation unit in vicinity of breached dam segment. **c** Dam Section A in Main Embankment. **d** Dam Section D in Perimeter Embankment

(a) Within the dam breach zone

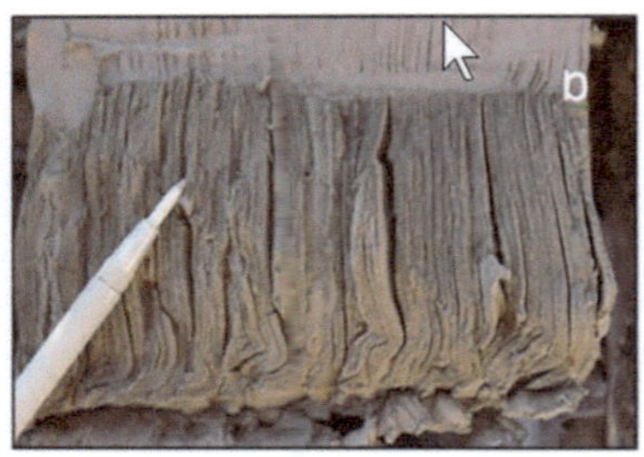

(b) Outside the dam breach zone

Fig. 3.99 Soil samples retrieved from weak Upper Glacio-Lacustrine foundation unit (from https://www.mountpolleyreviewpanel.ca/final-report). **a** Within the dam breach zone. **b** Outside the dam breach zone

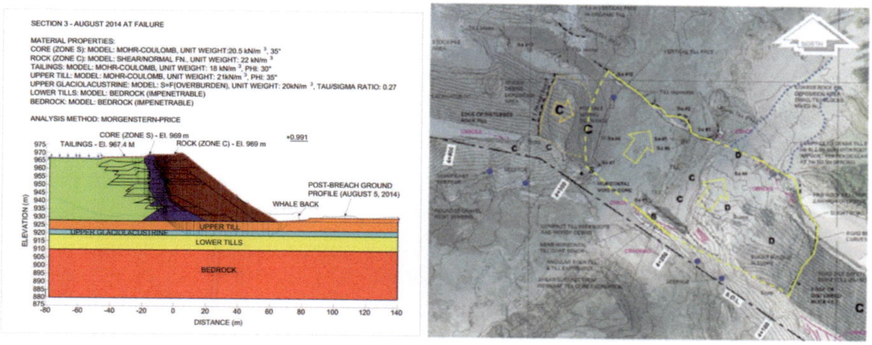

(a) Back analysis of dam stability at failure

(b) Inferred direction and extent of dam movement

Fig. 3.100 Dam stability and movement at time of failure (from https://www.mountpolleyreviewpanel.ca/final-report). **a** Back analysis of dam stability at failure. **b** Inferred direction and extent of dam movement

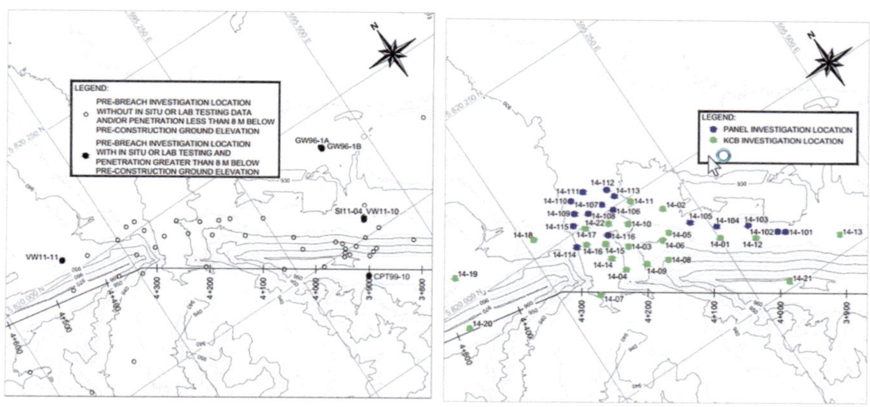

(a) Investigations prior to dam failure

(b) Investigations post-dam-failure

Fig. 3.101 Locations of Perimeter Embankment dam investigation drill holes—Station 3 + 800 to 4 + 500 (from https://www.mountpolleyreviewpanel.ca/final-report). **a** Investigations prior to dam failure. **b** Investigations post-dam-failure

$= \tau/\sigma = 0.27$ for the Upper Glacio-Lacustrine foundation unit. Figure 3.100b shows the inferred direction and extent of dam movement during the dam failure process.

Figure 3.101 shows the locations of dam foundation investigation drill holes for the Perimeter Embankment in the area between Stations 3 + 800 and 4 + 500, note the breach zone is between Stations 4 + 200 and 4 + 300. Holes drilled prior to the dam failure are shown in Fig. 3.101a, while those carried out after the failure are shown in Fig. 3.101b. As shown in Fig. 3.101a considerable number of holes were drilled during the dam design and construction phase before dam failure. Unfortunately, only 6 of these holes were drilled deeper than the critical depth of 8 m with insitu or laboratory testing program carried out. This shortcoming in the investigation programs at the design/construction stages sowed the seed of eventual dam failure during 2014 dam-raise construction.

Key Findings by the Review Panel

- The original foundation investigation programs did not identify the presence of a weak layer—Upper Glacio-Lacustrine unit under the breached dam segment.
- The shortage of embankfill construction materials persists over the dam-raising years. To cope with the fill shortage problem, the downstream slope of the failed dam section has been kept at a steep slope equal to the angle of repose, ~ 1.3 H: 1 V, rather than the flatter design slope of 2 H: 1 V.
- As the dam height increased to ~ 41 m at the breached segment in July/August, 2014, the moderately over-consolidated Glacio-Lacustrine soil becomes normally consolidated under the loading of the existing embankment fill. Thus, the incremental fill loading at the final dam-raise stage was applied under undrained shear condition that would induce positive excess pore pressure. The stability analysis using the undrained shear strength as shown on Fig. 3.100a indicated that the dam reached the stage of impending stability failure.
- Once this dam segment failed and moved downstream rapidly, water and tailings stored in the relatively full tailings impoundment overtopped the breached segment and downcut the embankment dam to the original dam foundation level.

The panel report contains a main text and 10 Appendices including a post-failure video taken from a helicopter in the morning of August 4, 2014 about 8 h after the breach initiation. The video was arranged by the Cariboo Regional District, showing the breached tailings impoundment and its downstream impacted areas.

The undrained shear failure of the breached dam segment can also be analyzed using effective shear strength of material along the potential failure surface using monitored pore pressure from piezometers installed at critical locations along the failure surface. Klohn et al. (1982) described a case history in British Columbia with similar dam foundation condition involving a Glacio-Lacustrine deposit. Dam stability was maintained during dam-raise construction by adopting the Observation Method, using the effective-stress stability analysis with prevailing foundation pore pressure monitored at critical potential failure surfaces. In fact, the original designer suspended ongoing dam-raise construction due to excessive foundation pore pressure build-up as well as foundation shear deformation in response to incremental increase of dam-fill loading. The timely suspension of construction most likely prevented the occurrence of a similar dam incident due to foundation failure (Burke & Smucha, 1979). Thus, tracking representative pore pressure development in foundation and damfill along potential failure surface is a prerequisite for the successful application of effective-stress approach to dam stability analysis (Klohn & Lo, 1978; Klohn et al., 1982).

Longer-Term Ramifications

After the completion of the Independent Expert Engineering Investigation and Review on January 30, 2015, there are two other ongoing investigations conducted, respectively, by the Chief Inspector of Mines and the Conservation Officer Service of the B.C. Government to run their courses. The Mount Polley Mine was allowed to reopen its mining operation in latter part of 2015, initially with the use of depleted Springer Pit for tailings storage. In the meantime, the mine is continuing its water quality monitoring program under the guidance of the Ministry of Environment. In March 2015, all mining companies with applications currently under British Columbia environmental assessment are required to consider the possibility of a tailings disaster, and to evaluate the environmental, health, social and economic impacts of such an accident.

The long-term impact of the breach of the Mount Polley Tailings Dam on Likely, a rural community with a population of about 350 and its surrounding area remains to be closely monitored. The community is situated on Quesnel Lake in the foothills of the Cariboo Mountains. The Quesnel Lake is the deepest fjord-type lake in North America with an inland temperate rain forest. The area supports outdoor recreational activities such as snowmobiling, mountain biking, all-terrain-vehicle (ATV) trails, white water rafting, kayaking, fishing, hunting and outback camping. The gradual restoration of the environmental health of the area is the key for its recovery and long-term development.

Ten years later, Hosgood (2024) provided an update of the local conditions related to the breach of Mount Polley Tailings Dam. On December 10, 2024, the B.C. Conservation Officer Service, Environment and Climate Change Canada and Fisheries and Oceans Canada charged Imperial Metals under the federal Fisheries Act, and the first court hearing was scheduled on December 18, 2024 in the BC Supreme Court in Vancouver. The biggest local concern is the deterioration of Quesnel Lake such as: cloudy water, slimy shorelines, clogged filtration systems, and heavy metals from lake bottom sediment particularly during spring and fall.

Lessons Learned

It is unfortunately that the Mt. Polley tailings dam failed, because the local existence of a weak Glacio-Lacustrine unit was not recognized in time. It behooves us to remember Terzaghi's caution (1929) on the importance of minor geologic details.

The Mt. Polley Dam foundation failure led to the following improvements at the B.C. Ministry of Energy, Mines and Low Carbon Innovation:

- revision of Health, Safety and Reclamation Code for B.C. Mines in 2016;
- increase of mine inspectors to strengthen their inspection regime; and
- requirement of new mine application to assess the possibility of a tailings disaster, and to evaluate the environmental, health, social and economic impacts of such an accident.

3.17 Case 16 November 5, 2015 Samarco Fundão Tailings Impoundment Dam Static Liquefaction Flow Failure, Brazil

This case history is based mainly on the comprehensive investigation report by the Fundão Tailings Dam Review Panel (2016) and its appendices. The review panel consists of four internationally renowned members: Prof. Norbert R. Morgenstern (Chair), Messrs. Steven G. Vick, Cássio B. Viotti, and Bryan D. Watts.

The Samarco Fundão tailings impoundment, shown in Fig. 3.102, is located in the state of Minas Gerais (MG) in south east Brazil (see Fig. 3.113). The Samarco Mine is jointly owned by BHP Group and Vale SA. The Fundão tailings dam, emulating the existing Germano Buttress dam design, used two saprolite earthfill starter dikes: the upstream one to retain slimes and the downstream one to retain sand tailings as shown in Fig. 3.103. Work on the Fundão impoundment began in 2005. As shown in Fig. 3.102, the original tailings impoundment employed a decant system consisting of a main and secondary drainage gallery to remove excess supernatant water from the impoundment. Dam underdrainage facilities were provided to remove collected seepage and consolidation water from the tailings dam.

The key design requirement of the Fundão Dam was to maintain a well-drained sand tailings section (about 100 m high and 200 m wide) from the ground surface to the design ultimate crest at El. 920 m, as shown in Fig. 3.103. The sand tailings level was to be kept above the slimes level at all times. This design is based on the relative production rates of 70% tailings sand versus 30% slimes at the Samarco Mine, see Fig. 3.105 for the representative grain size curves of tailings sand and slimes. However, this production ratio could not be maintained during operation, resulting in difficulty in achieving the above key dam design premise. This unfortunate fact sowed the seed of the dam's eventual demise. No corrective actions had been taken over the seven long construction years from 2008 to 2015.

Significant Fundão Dam Incidents from 2009 and 2015

Construction of the Dike 1 Starter Dam, with its requisite drains and decant galleries, was completed in October, 2008, and full-scale discharge of sand tailings began on April 13, 2009. A series of dam incidents occurred from 2009 to 2015 due to various design and construction issues: involving piping internal erosion exiting at the downstream slope of the Starter Dam (see Fig. 3.104), settlement and sinkhole development on tailings surface above Main Gallery (see Fig. 3.106) and Secondary Gallery (see Fig. 3.107), daylighting of seepage near the left abutment at Els. 820 m, 845 m, 855 m, 860 m, 867 m, damfill cracking on left dam segment at El. 860 m and El. 875 m, daylighting of seepage near the right abutment at El. 855 m and El. 860 m, and movement of the as-built dam slope and seepage saturation and erosion of damfill on August 26, 2014. Stability back-analyses showed that the static safety factor of the downstream dam slope on August 26, 2014 was about 16% higher than unity, corresponding to the moment just prior to the dam failure on November 5, 2015. Figure 3.108 shows the as-built dam section at the critical left abutment based

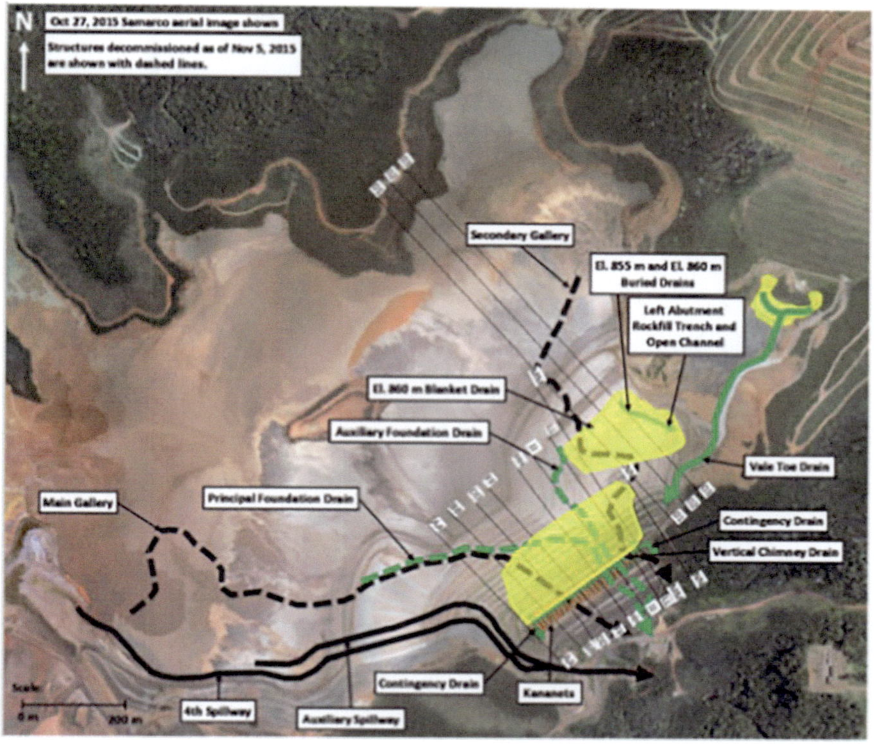

Fig. 3.102 General arrangement of the Fundão dam in November, 2015

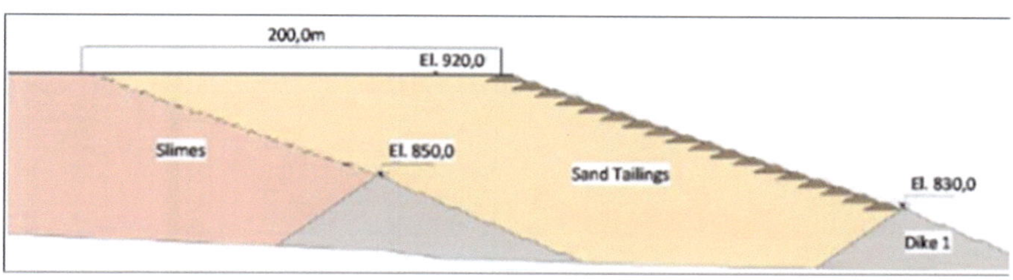

Fig. 3.103 Upstream raising of Fundão tailings dam. *Source* Fundão Tailings Dam Review Panel (2016)

on construction records with the presence of weak slimes near dam base, deviating substantially from the design section shown in Fig. 3.103. This margin of safety was lost over the next 14 months of ongoing dam-raise activities.

Dam Breach and Aftermath

In the afternoon of November 5, 2015, the Fundão Dam failed suddenly. Samarco construction personnel were at the damsite carrying out their routine work, as the dam

Fig. 3.104 Internal erosion of Starter Dam daylighted on downstream slope

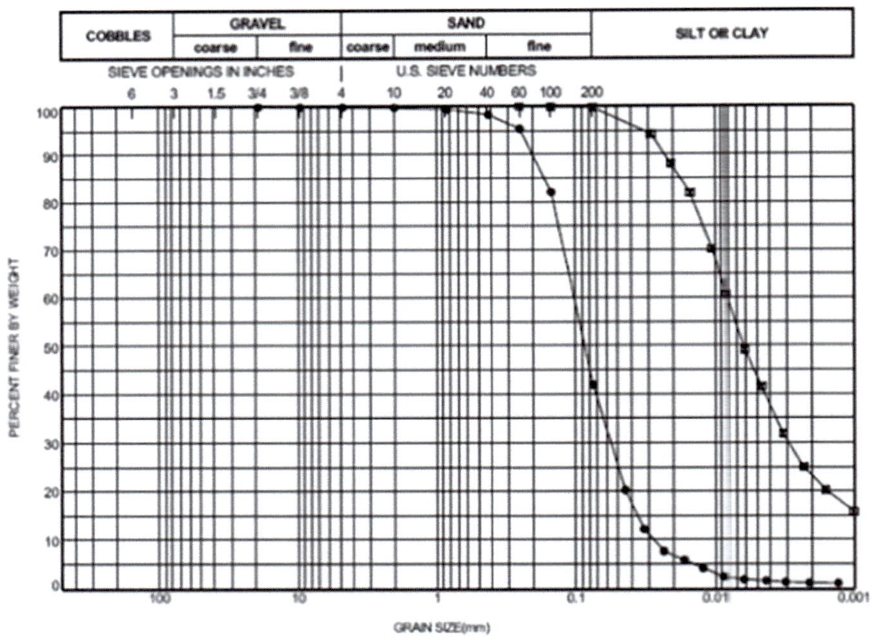

Fig. 3.105 Typical grain size curves for tailings sand and slimes

failure progressed rapidly in a matter of hours from showing initial signs of trouble to the final dam breach right in front of their eyes. Surviving eyewitnesses of this traumatic tragedy demonstrated their quick reflexes and tremendous speed. Unfortunately, some of their colleagues perished at the scene. The survivors preserved precious accounts of the failure progression, which are rarely available in most case histories of dam failure. Figure 3.109 shows the locations of these site crew members before dam failure, while Fig. 3.110 shows the same area with the dam

Fig. 3.106 Settled tailings surface with water vortex above defective main drainage gallery—July 9, 2010

Fig. 3.107 Sinkhole developed above defective joint of secondary drainage gallery—November 25, 2012

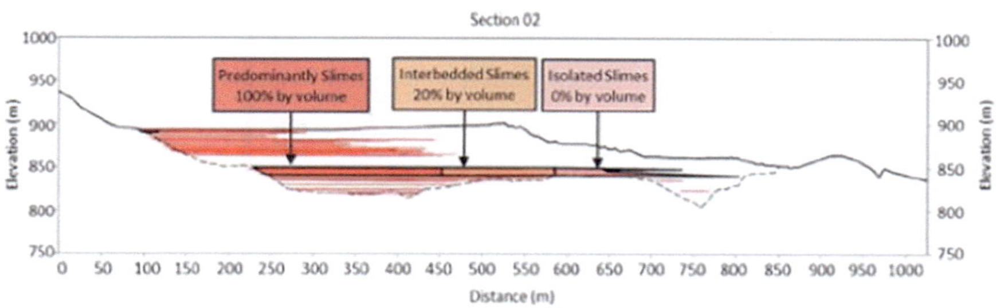

Fig. 3.108 Slime distribution at a critical dam section on left abutment

stripped away by the erosion of sand/slimes slurries to reveal, more or less, the original topography before dam construction. The ensuing mud flow engulfed nineteen lives. About 32 million cubic metre of tailings was lost, representing 61% of the impoundment content—an unusually high proportion based on statistics of tailings dam failure case histories.

Fig. 3.109 Dam failure observations in the afternoon of November 5, 2015—eyewitnesses' locations at dam failure

Fig. 3.110 Aerial view of Fundão dam and tailings impoundment after dam failure

Figure 3.111 shows the condition of the water course immediately downstream of the failed dam in Minas Gerais state, while Fig. 3.112 shows the contaminated Rio Doce flow discharging into the Atlantic Ocean at its mouth in Espírito Santo state. Figure 3.113 shows the locations of impacted sites along Rio Doce from Minas Gerais (MG) to Espírito Santo (ES) states.

"Our history was taken along with the mud. Paracatu and Bento Rodrigues will never be what they were …". Lamented Maria do Carmo Pereira Ramos, a former resident of Paracatu, by reading a poem (http://abcnews.go.com/International/wireStory/year-brazil-valley-waits-recovery-mine-flood-43123806). Her melancholy words probably expressed the heartfelt sentiment of all those people whose lives have been rudely disrupted by this and other tragic events of tailings dam failure.

Fig. 3.111 Inundation at Bento Rodrigues, Mariana, Minas Gerais. *Source* Rogério Alves/TV Senado

Fig. 3.112 Contaminated Rio Doce water flows into the Atlantic Ocean—November 30, 2015. *Source* earthobservatory.nasa.gov

Lessons Learned

The main problem at the Fundão Tailings Impoundment is the fact that the tailings dam design is based on constructing a sand dam as shown on Fig. 3.103. However, during mining operation, the mine could not achieve this fundamental design premise. No effort was made to check and confirm that slimes contamination of the sand dam did not occur. Instead, the entire dam raise program had been plagued by various other issues, while the most critical issue of slimes contamination persisted as indicated on Fig. 3.108 by the review panel investigation.

The vivid recount of survival eyewitnesses, whose locations on the dam site depicted on Fig. 3.109, of the final demise of the Fundão Tailings Impoundment

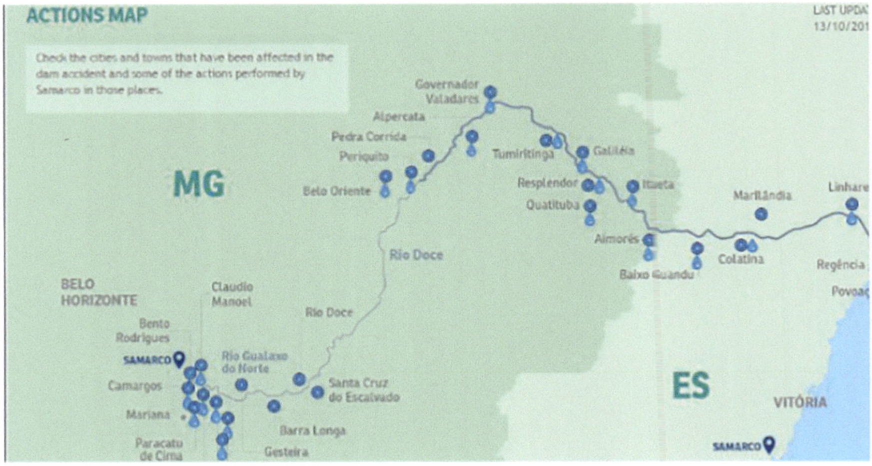

Fig. 3.113 Locations of downstream impacted sites along Rio Doce. *Source* Samarco

indicates that the dam collapse was progressing in a amazing speed. In a matter of hours, the vast solid-looking earth dam shown in Fig. 3.109 disintegrated and flowed away completely and exposed more or less the original topography within the impoundment footprint shown in Fig. 3.110. Unfortunately, the tragic scene would recur about three years later at another dam owned by the same mining company as described in Case 18. Only, this time the dam collapse process was captured by the dam-safety monitoring video cameras. Reuters (2019) indicated that the final settlement cost for the Fundão dam collapse could reach US $19 billion.

3.18 Case 17 March 9, 2018 Cadia Valley Northern Tailings Impoundment Downstream Slump of Southern Tailings Dam, Australia

Cadia Holdings Pty., a subsidiary of Newcrest Mining, is the owner and operator of Cadia Valley Operations that operated two tailings storage facilities: the Northern and Southern Facilities (see Fig. 3.114) near the town of Orange, New South Wales, Australia. On March 9, 2018, a slope slump occurred in the Southern Tailings Dam of the Northern Tailings Storage Facility (see Fig. 3.114b and 3.115), causing it to lose partial containment of tailings. Fortunately, the tailings were captured within the basin of the Southern Tailings Storage Facility. There were no injuries nor fatality due to a timely evacuation of personnel, following the mine's Dam Safety Emergency Plan.

The mining company engaged Ashurst Australia to co-ordinate and manage an independent technical investigation into the incident. An Independent Technical

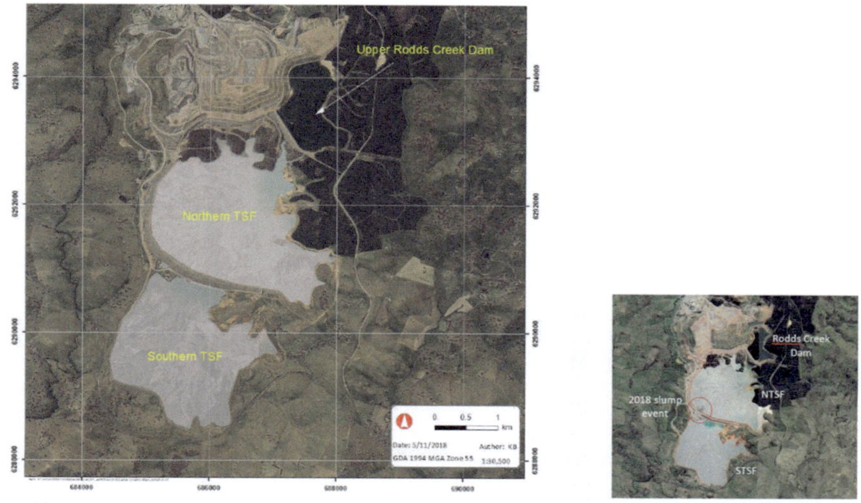

(a) General location map

(b) Location of March 2018 dam slump

Fig. 3.114 Location map of Cadia Valley Operation's Northern and Southern Tailings Storage Facilities (ITRB, 2019). **a** General location map. **b** Location of March 2018 dam slump

Fig. 3.115 Bird's-eye view of 2018 Cadia Valley Northern Tailings Impoundment Downstream Slump of Southern Tailings Dam (Hambrett, 2021, supplied: @farmerfromdownunder)

Review Board (ITRB) was established to determine the immediate technical cause(s) of the incident with assistance from an investigation team.

Cadia Valley Tailings Impoundments

As shown in Fig. 3.114, the Northern and Southern Tailings Storage Facilities are built across the Rodds Creek Valley. The Starter Dam of the upstream Northern Tailings Facility was built from 1998 to November 2001, while the Starter Dam of the downstream Southern Tailings Facility was completed in mid-2007. Figure 3.116 shows a typical cross section of the Northern Tailings Dam. The dam had been constructed in 10 stages. Stage 1 is the Starter Dam designed by Knight and Piesold,

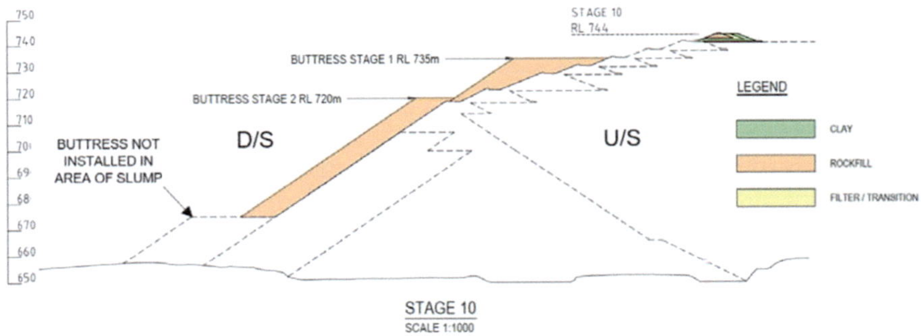

Fig. 3.116 Stage 10 construction of Northern Tailings Dam—addition of downstream berms Stages 1 and 2 (year 2017–2018)

with a crest elevation at El. 700 m. During construction of the Starter Dam, water in the Rodds Creek was diverted through a 1350 mm diameter steel conduit laid on the dam foundation. The diversion conduit was subsequently plugged by pumping low-strength material to fill the pipe.

Starting from Stage 2, design of Cadia Valley Tailings Facilities has been managed by Woodward Clyde and its successors from URS to AECOM. Up to Stage 3 with a crest elevation at El. 718.5 m in November 2005 the dam was raised as a conventional earthfill embankment dam. From Stage 4 to Stage 10, the dam was raised by the upstream dam construction method using earthfill to construct a series of crest dams up to a crest elevation of 744 m in March 2018 (see Fig. 3.116).

Stage 10 construction included the addition of two downstream buttress berms (Stage 1 and Stage 2 also shown in Fig. 3.116), and was managed by ATC Williams.

Findings of Independent Technical Review

The Stage 10 dam construction commenced on February 27, 2017 and generally advanced from the northwest to the south and east. By the end of July 2017 dam construction in the vicinity of the slump area was complete. Dam movements monitored in 2017 by terrestrial-based techniques were small. The movements accelerated in early 2018 as revealed by satellite-based techniques, obtained after the March slope slump event. Figure 3.117 shows the Southern Tailings Dam of the Northern Tailings Facility in the vicinity of the March 2018 slope slump in plan (Fig. 3.117a) and cross-section views (Fig. 3.117b). Figure 3.118a shows the Google Earth imagery of the Northern Tailings Facility acquired on May 10th of 2018, while Fig. 3.118b, c show line of sight cumulative displacements at Reference Points 1 to 6 on the Southern Tailings Dam over the respective periods of January 2016 to March 2018, and October 2017 to March 2018. It is important to note that there were about 2 months of precursory warning signs that the displacement of Reference Points 1 to 4 began to accelerate in January 2018.

(a) Plan view

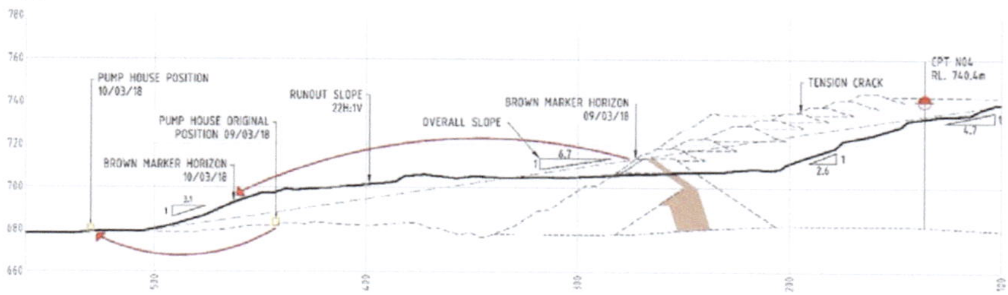

(b) Cross section through the slump area

Fig. 3.117 Southern Tailings Dam of Northern Tailings Storage Facility—plan and cross section through slope slump area. **a** Plan view. **b** Cross section through the slump area

(a) Google Earth imagery acquired 2018-5-10

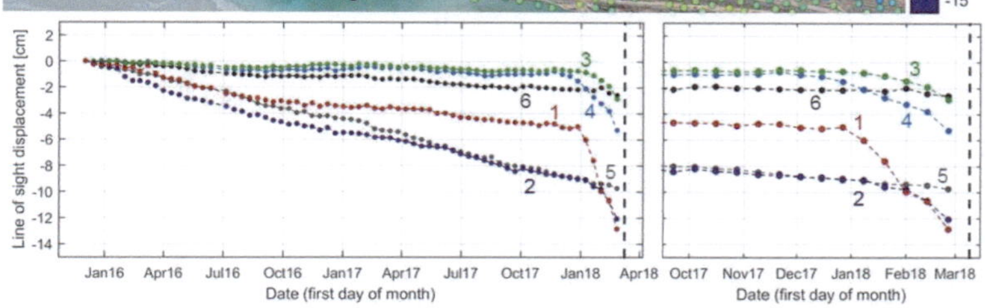

(b) Jan 2016 – Mar 2018 (c) Oct 2017 – Mar 2018

Fig. 3.118 Cumulative line-of-sight displacements at Cadia Southern Tailings Dam from InSAR persistent scatterer time series analysis of Sentinel-1 phase measurements. **a** Google Earth imagery acquired 2018-5-10. **b** Jan 2016–Mar 2018. **c** Oct 2017–Mar 2018

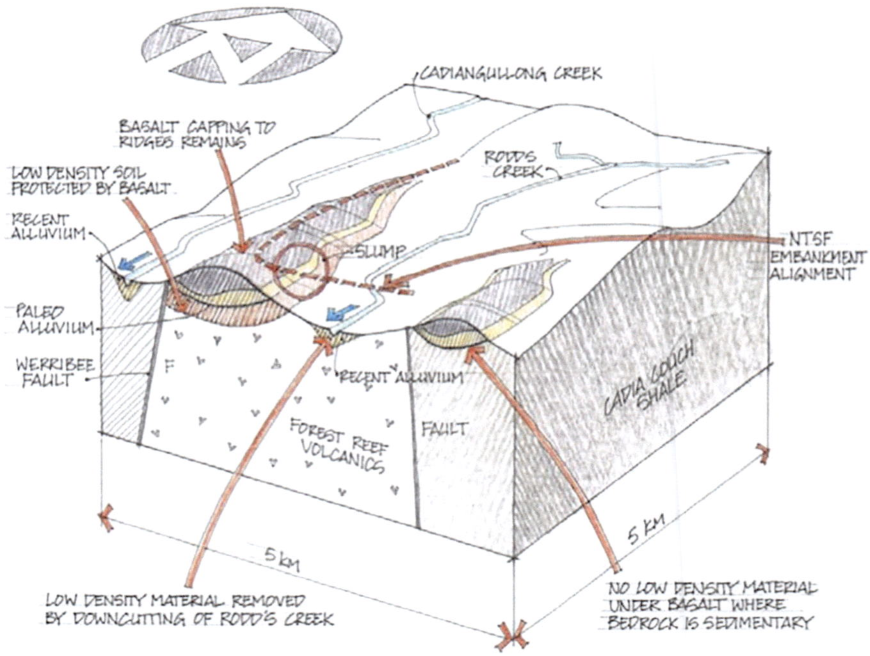

Fig. 3.119 Stream down cutting results in the formation of Rodds Creek

Weakness of South Tailings Dam

The foundation geology in the immediate vicinity of the South Tailings Dam is briefly discussed below with the aid of Fig. 3.119 illustration.

A major finding of the independent-review investigation is the existence of a low in-situ density foundation material within the Forest Reef Volcanics (FRV) unit. It is possible to define the low-density zone as FRV Unit A, and the rest as FRV Unit B. Figure 3.120 shows the plasticity chart of North Tailings Facility foundation materials including: FRV Unit A, FRV Unit B, Residual Basalt and Paleo Alluvium. The FRV Unit A material has a high void ratio, and brittle shear stress–strain relationship. Its drained and undrained shear strength are lower than that of the FRV Unit B. Thus, the FRV Unit A became a weak foundation layer, where horizontal foundation movement of the Southern Tailings Dam occurred.

Moreover, the tailings material also has a brittle shear stress–strain relationship under undrained loading condition as shown in Fig. 3.121. The slope stability and deformation analyses carried out by the review team (ITRB, 2019) demonstrated that concurrent with the increase of static shear loading due to the stage raises of the Southern Tailings Dam, the dam incurred ongoing downstream horizontal movement along the FRV Unit A until March 9, 2018. On March 9, the progressively reduced undrained shear strength of the tailings was no longer sufficient to

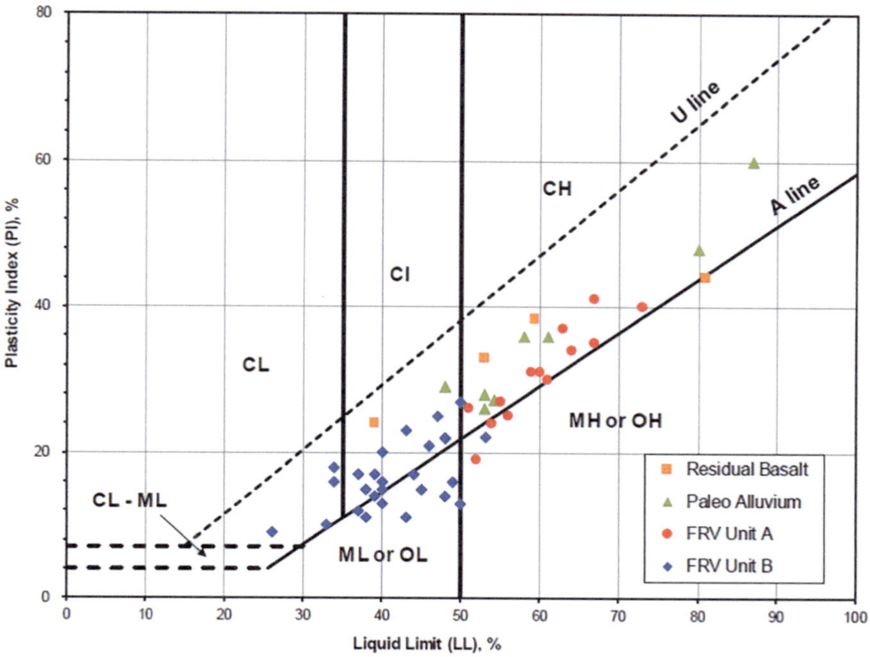

Fig. 3.120 Plasticity chart for Northern Tailings Storage Facility foundation materials

maintain the stability of the downstream slope, resulting a sudden slump. Thereafter, another slump involving liquefied tailings occurred on March 11. Eventually, following gradual dissipation of excess pore pressure and restoration of equilibrium of the post-slump dam geometry, the dam stabilized.

Macedo et al. (2024) refined the analysis carried out by ITRB in 2019. The refinement combined Lagrangian-based analyses with the Finite-Difference Method (FDM) to simulate small and moderate dam deformation up to March 9, 2018 with the Material Point Method (MPM) to track large dam deformation between March 9 and March 11, 2018. The former simulated initial slope slump triggered on March 9, while the latter tracked post-triggering dam deformation ending on March 11 involving interactions of both tailings and FRV Unit A foundation material in response to construction of buttress berms.

On the other hand, Elkhamra et al. (2023) back analyzed the slope slump using the simpler two-dimensional limit equilibrium approach and appropriately estimated mobilized shear strength for the FRV Unit A material and arrived at the conclusion that the original design strength for the foundation unit was overvalued.

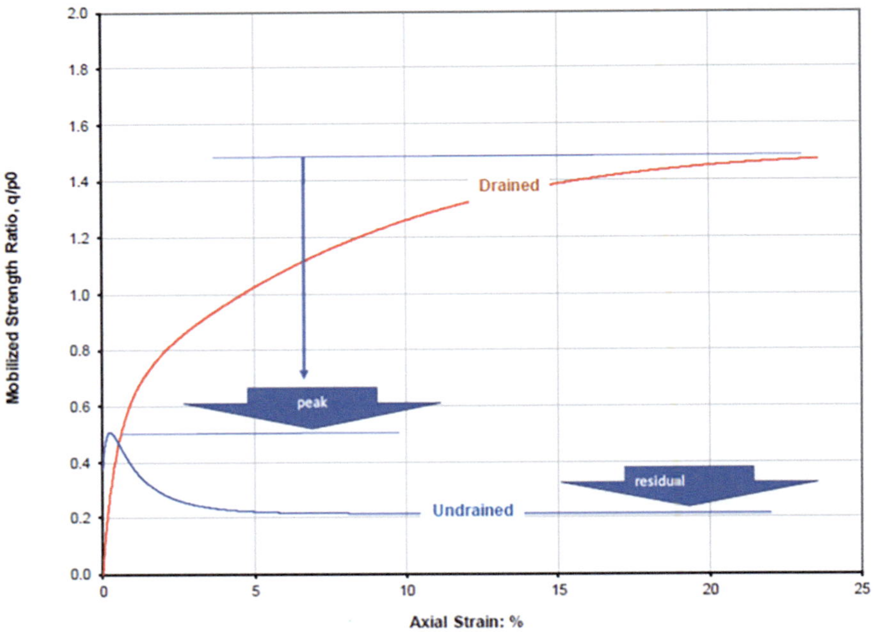

Fig. 3.121 Illustration of tailings shear stress–strain relations

Lessons Learned

- The consequence of downstream slope slump of the Southern Tailings Dam of the Northern Tailings Storage Facility is mitigated by the successful execution of the mining company's Dam Safety emergency plan.
- The designer of Cadia Tailings Storage Facilities has gone through two changes from 1998 to 2018. In addition, from 2000 to 2016, the second designer had gone through two company merges. These corporation changes on the overall operation of the tailings facilities might have some unknown unfavourable impact due to potential interruptions of continuity.
- It appears that the presence of a low-density weak FRV Unit A foundation layer as well as the vulnerability of stored tailings to static liquefaction might not have received deserved designers' attention to prevent the fate of eventual slope slump.
- Chua et al. (2019) described their use of Sentinel-1 radar images acquired from December 2, 2015 to June 25, 2018 at regular 12-day intervals and Sentinel-2 multispectral imagery acquired from July 2, 2017 to June 24, 2018 to study the downstream slope slump of the Southern Tailings Dam. They applied the Interferometric Synthetic Aperture Radar (InSAR) technique to measure surface displacements and interferometric coherence. The time series of surface displacements showed a significant increase in the rate of dam movement in the area that eventually slumped two months later. Their study of the Sentinel-2 multispectral

imagery data showed the timing, extent and effects of the slope slump as well as the rate of tailings movement.

- As discussed in Case 18, Das et al. (2024) applied an objective, probabilistic-learning algorithm to retrospectively analyze the risk of the 2019 Brumadinho Tailings Dam I. They indicated that the combination of spectral methods and second order statistical properties of InSAR displacement time-series data can reveal signs of a transition into an unstable deformation regime, and can provide sufficient early-warning that could help mitigate catastrophic slope failures. It would be useful to apply their method to the case of Cadia Valley South Tailings Dam to extend the method's application.
- The independent review board's investigation in 2018–2019 and further researches in 2024 by others shed more lights to this case history. It is reassuring that simpler limit equilibrium analysis approach (Elkhamra et al., 2023), when used appropriately, can provide sound and safe basis for dam design purposes. Moreover, sophisticated analysis methods (Macedo et al. 2024), in special conditions when requisite parameters are carefully evaluated, can be used to extend our understanding of complex mechanisms involved in a case history.

3.19 Case 18 January 25, 2019 Feijão Tailings Dam I Static Liquefaction Flow Failure, Brazil

Córrego do Feijão Iron Ore Mine's (a subsidiary of Vale SA) Tailings Dam I is located 9 km northeast of Brumadinho, in the state of Minas Gerais, Brazil (see Fig. 3.122). The dam was about 83 m high. Two satellite photos of Dam I are shown on Figs. 3.123 (with a small pond) and 3.124 (with a much larger supernatant pond).

Dam I Liquefaction Failure Progress Captured on Videos

At 12:28 p.m. local time on January 25, 2019, the dam collapsed suddenly (see Fig. 3.125). The ensuing tailings debris mudflow rushed downstream for 8 km, burying offices, homes and killing 270 people (more than half were mine employees). Approximately 9.7 million cubic meters (Mm^3) of material were involved in the mudflow. This volume represents approximately 75% of total volume of about 12.7 Mm^3 of tailings and damfill. Figure 3.126 shows schematic presentation of Dam I cross section constructed by the upstream dam method (Fig. 3.126a), rapid progression of liquefaction flow failure of the tailings impoundment (Fig. 3.126b) and the downstream areas impacted by the tailings debris flow (Fig. 3.126c). Figure 3.127 shows a highway buried by the tailings mud in its low-lying segment. Figure 3.128 shows the distribution map of tailings dams constructed by the upstream dam method in Brazil.

The Feijão Tailings Dam I liquefaction flow failure was unusual, because it left a video record of the entire failure process [see Liber Media Brasil (2019), a Portuguese broadcast]. For a brief period of about 6 min, the dam was transformed from an intact hydraulic structure retaining tailings deposits and a supernatant pond to a raging

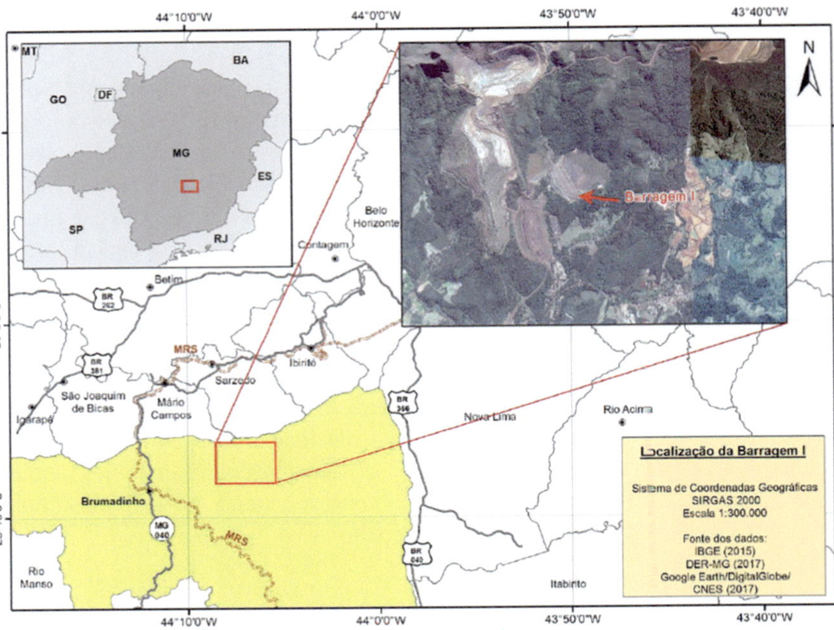

Fig. 3.122 Location of Dam I in south Brazil (from Tuv Sud, 2017)

Fig. 3.123 Aerial view of Dam I in foreground and Dam VI in background (from Brumadinho Engineering History, 2020)

debris flow current carrying the disintegrated dam and most of its stored content (see Fig. 3.129h). Perhaps due to the rapid speed of the dam breach, no warning sirens were set off resulting in fewer survivors to escape the disaster.

The Review Panel (2019) studied the video images: eight frontal views of the downstream dam face (see Fig. 3.129a–h); and two back views of the dam from the

Fig. 3.124 Feijão Tailings Dam I in foreground (from Brumadinho Engineering History, 2020)

Fig. 3.125 Failed Feijão Tailings Dam I in background (from Brumadinho Engineering History, 2020)

tailings impoundment (see Fig. 3.129i, j). As shown on Fig. 3.129, it only took ~ 6 s from noticing some visible sign of crest movement in Fig. 3.129a to the well defined settlement of the downstream slope in Fig. 3.129d. For the next 12 s, the dam outward movement from being merely noticeable in Fig. 3.129e progressed to the vigorous debris flow in Fig. 3.129g. All told, from start to finish it only took ~ 6 min for the dam to collapse.

Feijão Tailings Impoundment I

Feijão Tailings Impoundment I was originally owned and operated by a German mining company Ferteco until April 27, 2001, when Vale acquired Ferteco. Tailings Dam I was constructed in a valley underlain by banded gneisses bedrock, which is overlain by weathered saprolite, residual, and lateritic colluvial soils. A natural

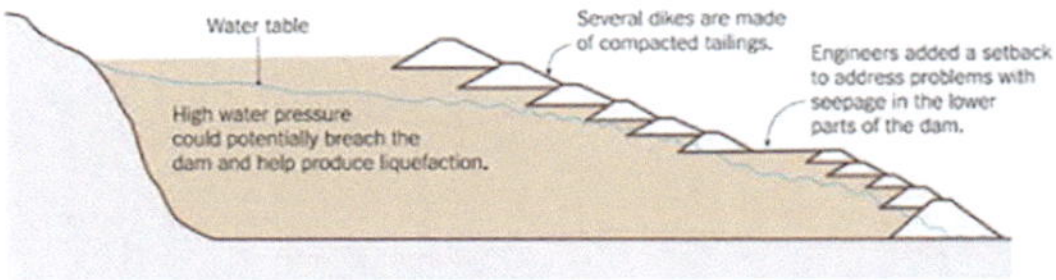

(a) Cross Section of Dam I constructed by the upstream dam method

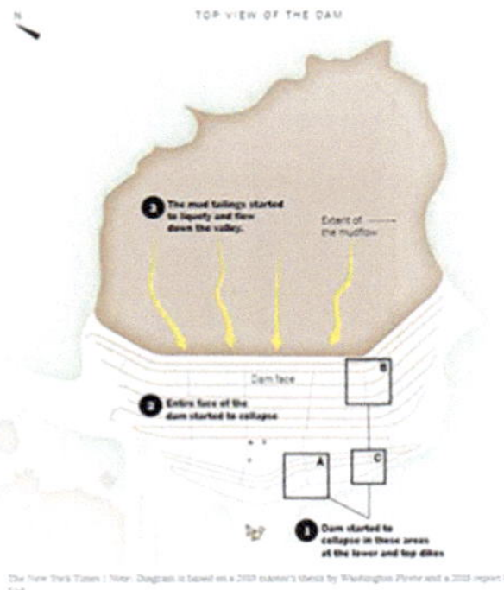

(b) Rapid progression of dam failure by liquefaction

(c) Areas impacted by the dam-collapse mudflow

Fig. 3.126 Schematic presentation of Dam I flow failure (from NY Times, 2019). **a** Cross section of Dam I constructed by the upstream dam method. **b** Rapid progression of dam failure by liquefaction. **c** Areas impacted by the dam-collapse mudflow

Fig. 3.127 A highway buried by the mudflow in its low-lying segment (Antonio Lacerda/EPA, via Shutterstock, from NY Times, 2019)

creek running along the talweg of the valley was blocked by the tailings dam. Dam I was constructed from 1976 to 2013 with 10 raises from the Starter Dam (see Fig. 3.130). During the 37 construction years, there were 5 consultants and even more numerous construction companies involved. The dam had limited internal drainage features, such as chimney drains, drainage blankets and sub-horizontal drainage pipes, installed in peripheral dykes along the downstream slope (see Fig. 3.130).

The tragic failure of Samarco Fundão tailings dam in November 2015 (see Case 16) had profound impact on the management of Vale SA, who is the joint owner of that dam. Dam safety became the highest priority of its operation. As it implemented

Fig. 3.128 Distribution map of upstream tailings dams in Brazil (by Scott Reinhard I *Source* Agenda Nacional de Mineração, from NY Times, 2019)

this focus on Feijão Tailings Impoundment I, it began a systematic program to reduce safety risks of Dam I. It started to decommission Dam I Impoundment, and dewater Dam I. Unfortunately, although the goal was correct; yet its implementation was still fraught with inherent risks already embedded in the as-built dam constructed by the Upstream Dam Method. The intention was to re-process the tailings deposits for recovering the economic value of its residual iron content, after the impoundment is stabilized with various remedial measures.

From February to June 2018, in an effort to drawdown pore pressure within the saturated damfill, the mine installed 15 deep horizontal drains, i.e., perforated PVC drainage pipes DHP01 to DHP15 as shown on Fig. 3.131. The typical length of the installed DHPs was about 60 m with the maximum length of 81 m. Drilling through the peripheral dykes was achieved with an air pressure of up to 600 kPa, and with additional water pressure of up to 400 kPa after advancing through the tailings. The installation of DHP15 caused internal disturbance of the damfill, loss of drill rod and loss of fines and seepage. DHP15 hole was grouted up, and sandbags were used to control local seepage. DHP06 was also abandoned. The rate of flow from these drains varied, and also decreased with time. By November 2018, only 8 of these drains were flowing at a rate ranging from 0.18 to 15 L/min.

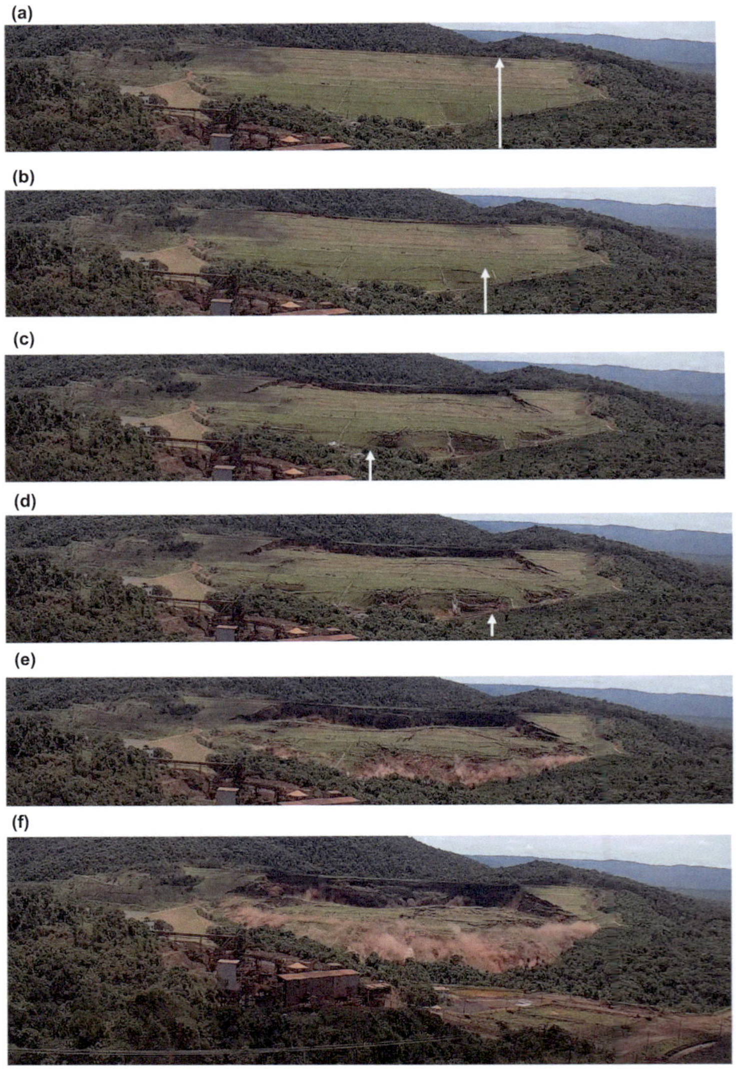

Fig. 3.129 **a** Dam I downstream face—2019 Jan 25 initial deformation near central part of crest (from Expert Panel, 2019). **b** Dam I downstream face—2019 Jan 25 bulging above dam toe (0.2 s after (**a**), from Expert Panel, 2019). **c** Dam I downstream face—2019 Jan 25 continued crest settlement and bulging above toe (5.5 s after (**a**), from Expert Panel, 2019). **d** Dam I downstream face—2019 Jan 25 extensive crest settlement and bulging above toe (5.8 s after (**a**), from Expert Panel, 2019). **e** Dam I downstream face—2019 Jan 25 fully developed slope failure moving out beyond toe (6.7 s after (**a**), from Expert Panel, 2019). **f** Dam I downstream face—2019 Jan 25 dam disintegrated rapidly (11 s after (**a**), from Expert Panel, 2019). **g** Dam I downstream face—2019 Jan 25 disintegrated dam began flowing out (18 s after (**a**), from Expert Panel, 2019). **h** Dam I downstream face—2019 Jan 25 dam breach practically completed (6 min and 25 s after (**a**), from Expert Panel, 2019). **i** Back view of Dam I from tailings impoundment—2019 Jan 25 dam moving forward away from pond (1 min after (**a**), from Expert Panel, 2019). **j** Back view of Dam I from tailings impoundment—2019 Jan 25 dam flowing forward (7 min and 25 s after (**a**), from Expert Panel, 2019)

(g)

(h)

(i)

(j)

Fig. 3.129 (continued)

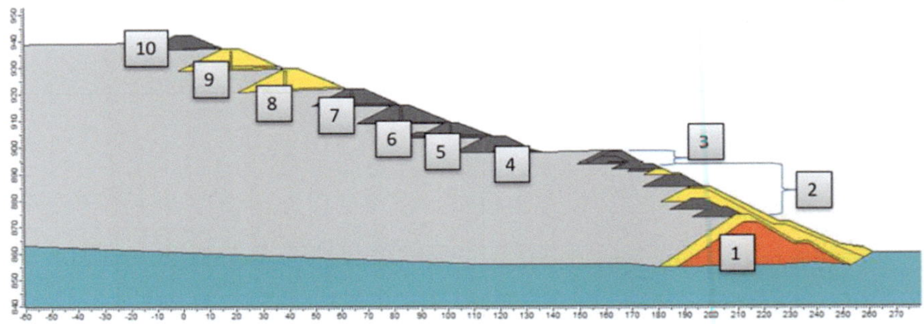

Fig. 3.130 Cross section of Dam I with 10 dam raises (from Expert Panel, 2019)

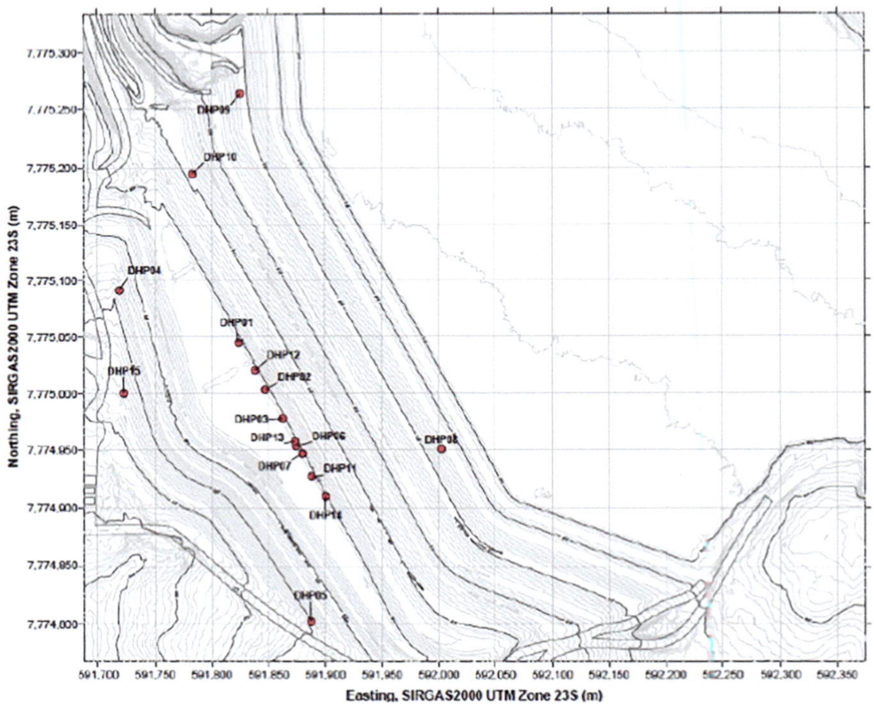

Fig. 3.131 Locations of deep horizontal drains installed in Feb–Jun 2018 (from Expert Panel, 2019)

From October 2018 to January 2019 the mine drilled 10 boreholes through Dam I for depths ranging from 28.5 to 76 m as shown on Fig. 3.132. These holes were drilled for purposes of planning for the decommissioning of Impoundment I. Drilling of Borehole B1-SM-13 was in progress when the dam collapsed on Jan 25, 2019.

Surface drainage ditches as well as 3 decant towers along the right dam abutment were used to remove excess supernatant water, surface runoff and any groundwater

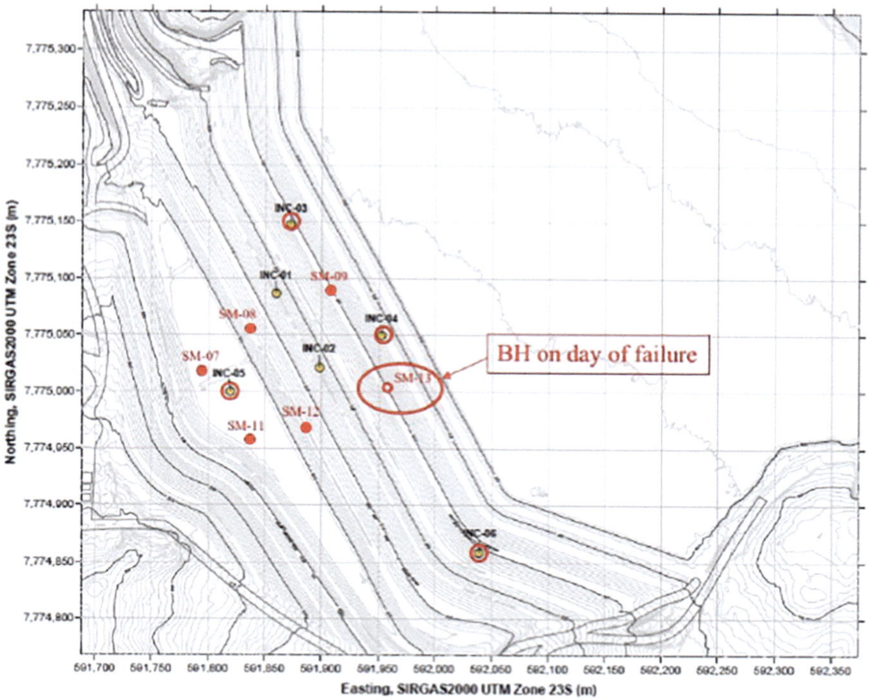

Fig. 3.132 Locations of boreholes drilled in Oct 2018–Jan 2019 (from Expert Panel, 2019)

discharge from the impoundment surface to Impoundment VI located to the right of the impoundment (see Fig. 3.124). Seepage and rainfall runoff on the dam downstream slope was collected by a series of concrete-lined channels. Flows in these channels were routed down the face of the dam and drained to the creek at the base of the dam. After July 2016, no tailings was deposited in the impoundment. To maintain a low water level of the supernatant pond, a pump and piping system was installed in mid-2016 (see Fig. 3.136). A backup pump was later added to the system in case the primary pump failed.

Expert Panel Investigation (From Expert Panel, 2019)

Vale commissioned an investigation on the technical causes of Tailings Dam I failure by an international expert panel consisted of Dr. Peter K. Robertson (Chair), Dr. Lucas de Melo, Dr. David Williams and Dr. G. Ward Wilson. The Panel with its consultants reviewed historical data and documents, conducted additional field and laboratory testing to confirm and supplement existing data, and engaged in computer modeling of the dam performance. The panel issued a main report and 10 appendices (Expert Panel, 2019), which forms the main basis of the following section.

Figure 3.133 shows the cross section of Dam I near the mid-crest with the distribution of coarse, fine tailings and slimes. Figure 3.134 shows the range of tailings gradation curves from historical data together with individual gradation curves from samples retrieved by the Panel. The Panel found that the tailings has high iron content, which contributes to light bonding at grain contacts. The specific gravity is 4.9 for the coarse tailings and 4.0 for the slimes.

The climate regime near Dam I shows distinct wet and dry seasonal rainfall periods. Moreover, long-term fluctuations of the annual rainfall were superimposed on these seasonal variations. To study the influence of rainfall on the soil suction, an infiltration analysis was carried out using rainfall data from January 2016 to January 2019. Figure 3.135 shows the pore pressure profile at a location near the dam crest at the mid-crest section. The computed suction pressure in the unsaturated zone decreases significantly as rainfall infiltrates the zone. The loss in suction associated

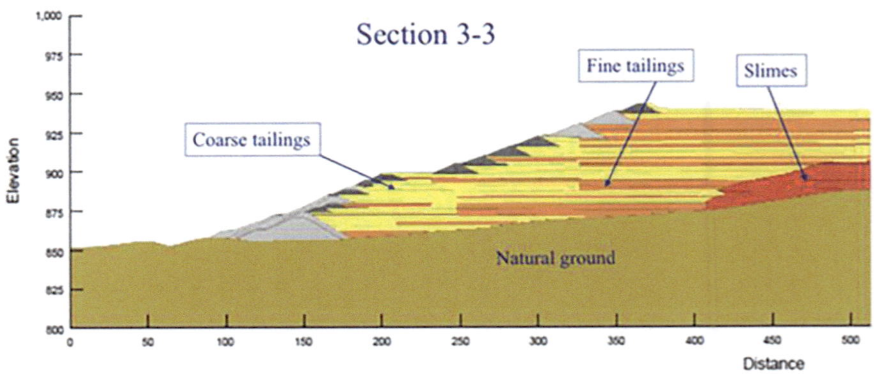

Fig. 3.133 Cross section of Dam I near mid-crest (from Expert Panel, 2019)

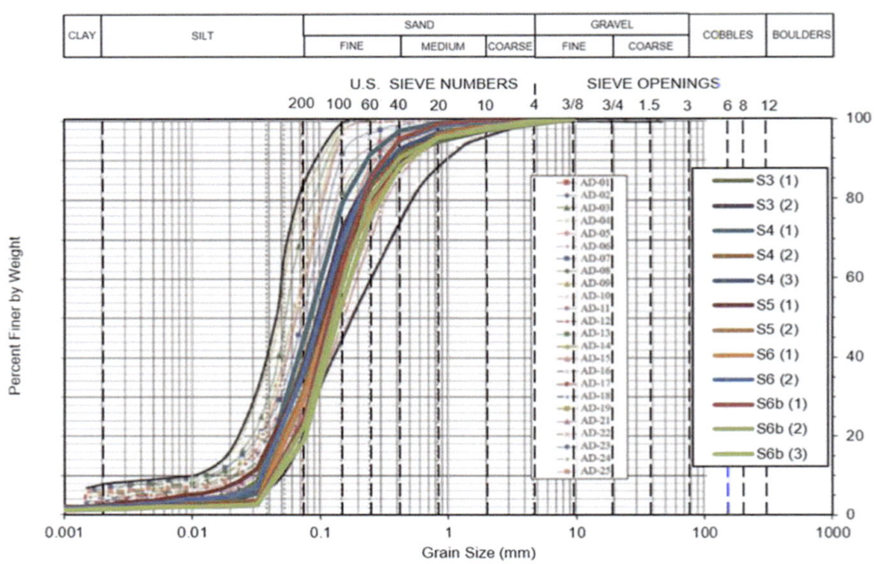

Fig. 3.134 Gradation curves of tailings (from Expert Panel, 2019)

with infiltration is in the order of 50 kPa, which in turn reduces the undrained shear strength of this zone by about 10–15 kPa. The figure also illustrates that the average water pressure below the water level can be approximated by a 50% hydrostatic pressure profile.

The Panel carried out simulated seepage analysis through the dam and calibrated pore water pressure with those measured values obtained from installed piezometers. Figure 3.137 shows calculated pore water pressures in a dam section near the mid-crest. The Panel also carried out special laboratory tests to study tailings shear stress–strain characteristics. Figure 3.138 shows a pair of anisotropically consolidated and

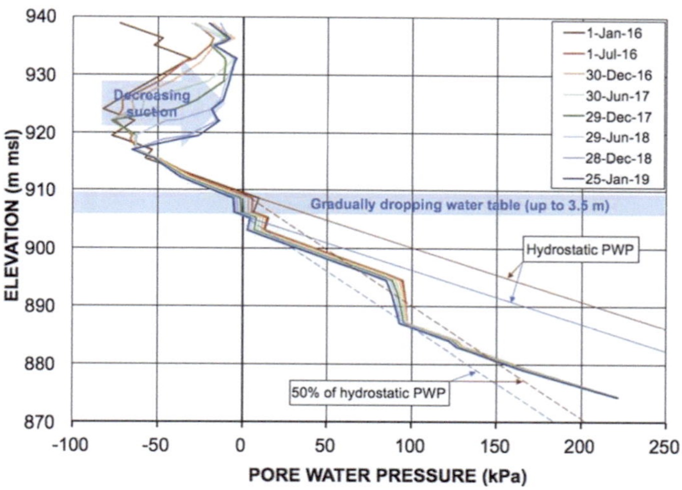

Fig. 3.135 Pore pressure profile obtained from an infiltration simulation analysis for a point located near mid-crest (from Expert Panel, 2019)

Fig. 3.136 Pumping system to drain supernatant pond water to Impoundment VI via a pipe and decant tower (from Expert Panel, 2019)

undrained-sheared triaxial shear tests. The test on the left was sheared under strain-controlled condition. The stress–strain curve shows strain-softening behaviour i.e., the strength reaches a peak value at a small strain then drops off rapidly to a residual value. The test on the right was tested under stress-controlled condition. Thus, once the strength reaches its peak value, the sample failed suddenly under the applied dead load.

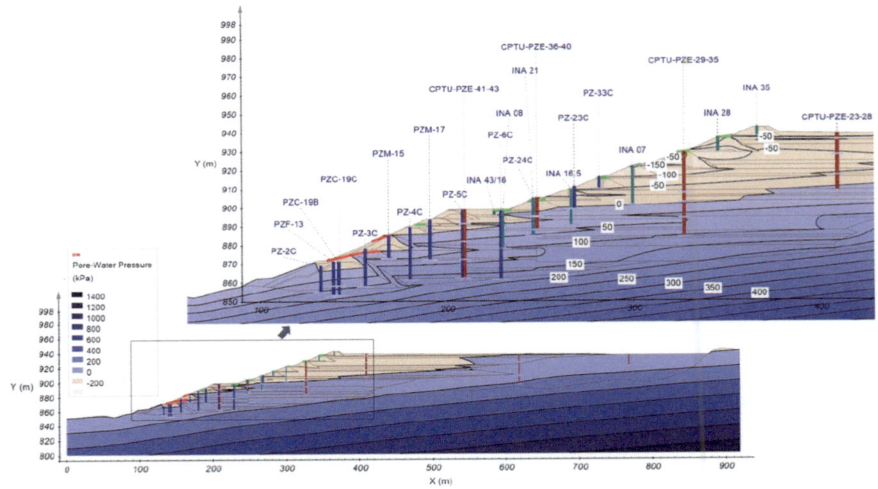

Fig. 3.137 Simulated seepage through Dam I near mid-crest section (from Expert Panel, 2019)

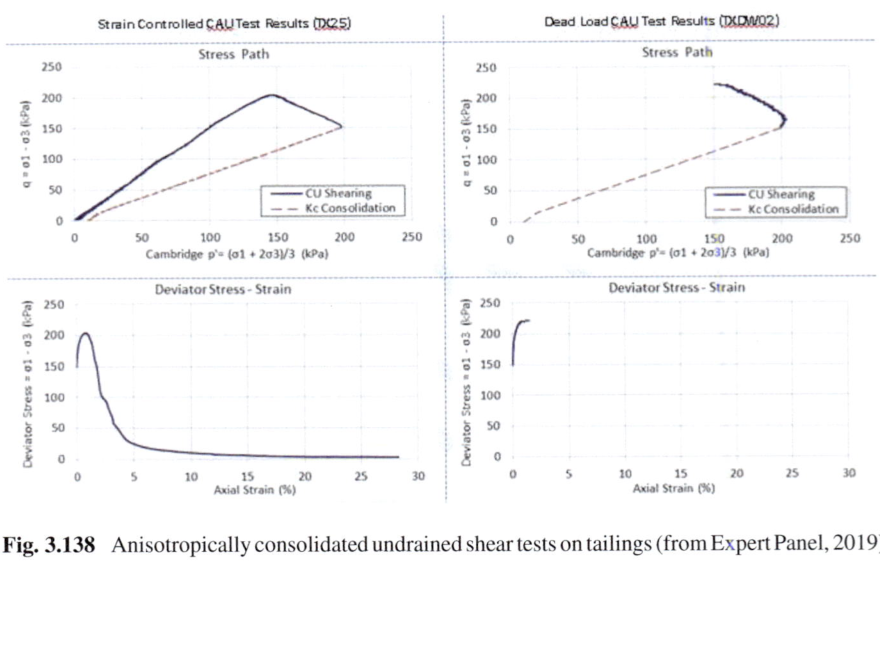

Fig. 3.138 Anisotropically consolidated undrained shear tests on tailings (from Expert Panel, 2019)

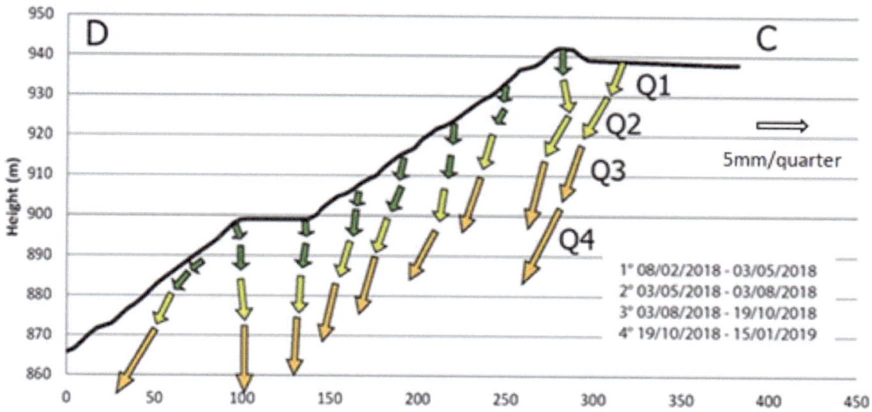

Fig. 3.139 Deformation vectors of Dam I surface for 4 quarters prior to dam failure (Expert Panel, 2019)

The Panel reviewed Dam I deformation data for the period prior to dam failure, including survey on prisms installed along the crest, inclinometers, ground-based radar data and satellite (InSAR) data. It appears that only the satellite (InSAR) data provides useful information. Figure 3.139 shows deformation vectors of Dam I surface for 4 quarters from February 2018 to January 2019 obtained from the InSAR data in the central region of the dam. These vectors show mainly downward with general outward horizontal movement. In addition, the observed movements for each quarter increase with time until final dam failure due to static liquefaction (Rodríguez-Pacheco, 2019). Holden et al. (2020) used both high and low resolution imagery from 4 sets of InSAR data including: 2 from Sentinel-1 and 1 each from TerraSAR-X and COSMO-SkyMed. They used data for the period of 2 years up to the failure of Dam I, and evaluated the statistical significance of the measured signals.

The Expert Panel carried out both 2-dimensional and 3-dimensional deformation and stability analyses to simulate Dam I performance up to the point of impending collapse. Figure 3.140 shows the calculated horizontal displacement of Dam I when static liquefaction was triggered by the ongoing dam creep movement after tailings deposition ended in July 2016 and the loss of shear strength of 15 kPa in the unsaturated zone, due to its saturation by rainfall infiltration.

Figure 3.141 from the Panel report Appendix I shows striking similarities of dynamic signatures of seismographs recorded at the FJAO seismograph station, located ~ 2.4 km from Dam I, during two important events at the dam. The figure on the left corresponds to the dam collapse on January 25, 2019, while the one on the right corresponds to an emergency incident during drilling of deep horizontal drain DHP15 on June 18, 2018 (about 7 months prior to the eventual dam failure). The drilling incident of the ~ 76 m deep hole caused excess pore pressure in the vicinity of the borehole and mudflows resulting loss of damfill material and water in the local internal drainage system. Measures taking in this emergency period of about 3 days including: containing mudflows by sandbags, and adjusting outlets

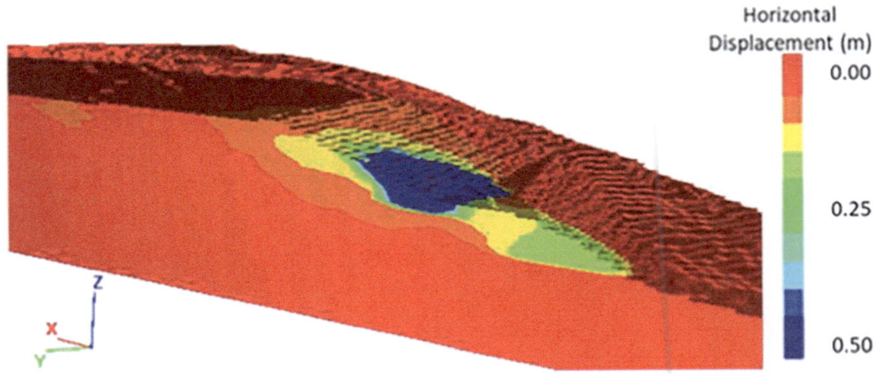

Fig. 3.140 Calculated horizontal deformation of Dam I from a 3-dimensional numerical model (from Expert Panel, 2019)

of internal drainage system in the impacted region to control excess pore pressure. After the mudflow and excess pore pressure problem gradually attenuated, no more installation of deep horizontal drains were attempted. However, additional drilling activities had taken place for ongoing instrumentation installation and documentation purposes from October 2018 to January 25, 2019, the day when Dam I collapsed suddenly.

Circumstances of Sudden Dam Failure

To further clarify the circumstances of the apparently sudden failure of Dam I, Vale and the Federal Public Prosecutor's Office (MPF) reached an agreement to perform specialized computational analyses of Dam I. Under this agreement, International

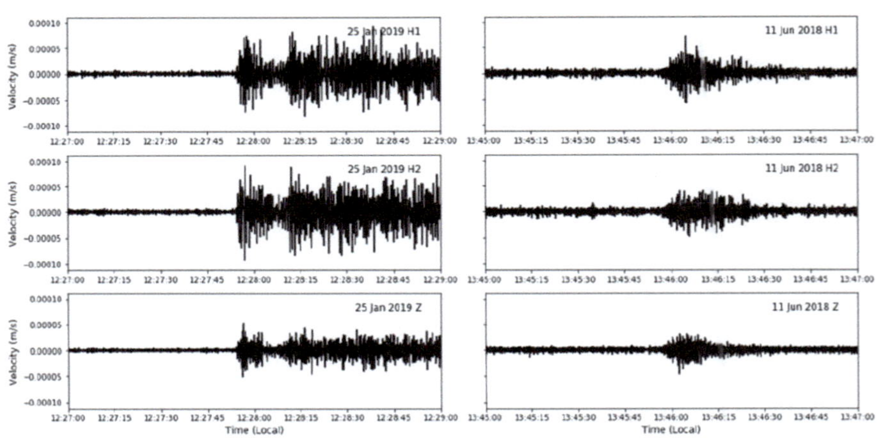

Fig. 3.141 Dynamic signatures at station FJAO of the initiation of Dam I failure on 25/01/2019 and similar signatures of the localised liquefaction failure at DHP15 on 11/06/2018 (EPR, App I, Figure 14)—H1 (FJAO horizontal 1), H2 (FJAO horizontal 2), Z (FJAO vertical)

Centre for Numerical Methods in Engineering (CIMNE) was tasked to carry out the analyses. The work had been directed by Drs. Marcos Arroyo and Antonio Gens (CIMNE, 2021).

CIMNE reviewed databases available in both Vale and MPF files, conducted additional field investigations and laboratory testings and carried out visco-plastic, effective-stress numerical modelling of Dam I. A suite of numerical analyses has been performed to assist in the interpretation of the failure circumstances. The analyses incorporate available information on dam geometry, construction history, material parameters and hydraulic/environmental conditions. The analyses led to the inferences of the following conclusions:

(1) the emergency incident occurred in June 2018 during installing deep horizontal drain DHP15 and the final collapse of Dam I on January 25, 2019 during drilling piezometer borehole BI-SM-13 involved spontaneous static liquefaction; and
(2) Dam I finally collapsed on January 25, 2019 apparently indicated that it had reached the point of no return.

Zhu et al. (2024) provided another numerical simulation of the evolution of the eventual failure of Dam I, following its history of staged construction, stoppage of tailings inflow and ongoing care and maintenance. Their simulation used the approach of tracking the growth of the slip surfaces until the formation of the merged critical final slip surface leading to the dam collapse.

Das et al. (2024) proposed an objective, probabilistic-learning algorithm to detect and characterize the risk of a slope failure, based on spectral analysis of serially correlated displacement time-series data. The algorithm is applied to satellite-based interferometric synthetic radar (InSAR) displacement time-series data to retrospectively analyse the risk of the 2019 Brumadinho Tailings Dam I. Two potential risk milestones were identified and signs of a definitive but emergent risk (from February 27, 2018 to August 26, 2018) and imminent risk of Dam I collapse (from June 27, 2018 to December 24, 2018) were detected by the algorithm as the empirical points of inflection and maximum, respectively on risk trajectory. Moreover, this precursory indication of risk of failure was detected as early as at least five months prior to the actual dam collapse on January 25, 2019. The results of this study demonstrate that the combination of spectral methods and second order statistical properties of InSAR displacement time-series data can reveal signs of a transition into an unstable deformation regime, and that this algorithm can provide sufficient early-warning that could help arrest the progress of development of a potentially catastrophic dam slope failure.

Lessons Learned

The collapse of Feijão Tailings Dam I is an important historical case history for the following reasons:

(1) The visual record of the collapse of Dam I probably is the first of such a record in the world. The speed of the demise of such a large hydraulic structure is only believable when we are confronted by the solid evidence of the video.

It appears that Dam I managed to survive with a small margin of safety over its life span. After 2016, tailings ceased to be added into the impoundment. The mine tried to reduce pore water pressure in the dam by installing 13 deep horizontal drains with some effect. Unfortunately, for an upstream tailings dam the bulk of the dam was constructed by hydraulic deposition of tailings slurry with layers of coarse, fine tailings and slimes distributed in an erratic manner throughout the impoundment. It was rather difficult to dewater such an earth structure effectively. Hence for tailings dams having similar characteristics like Feijão Dam I, all stakeholders should recognize that the risk of static liquefaction failure of this type of dam is very high, and all reasonable measures to safeguard lives and properties in the downstream vulnerable areas should be taken.

(2) CIMNE (2021) indicates that the numerical simulation analyses of circumstances of two events, i.e. June 2018 emergency suspension of installation of Deep Horizontal Drain DHP15 and January 25, 2019 sudden dam failure during installation of piezometer BI-SM-13 are similar in nature related to spontaneous static liquefaction. The outcomes of the two events differ in the former being stopped with only localized failure, while the latter progressing to the complete collapse of the dam in about 6 min.

The irony here is the fact that Dam I failed while Vale was attempting to improve its stability. We would never know whether the drilling activities in the last days of Dam I was indeed the last straw that broke the proverbial "camel's back". We do know, however, that one should never continue the flawed "Upstream-Construction-Method" relying on a flimsy margin of safety full of uncertainty.

(3) The unsuccessful attempt by Vale to reclaim and re-process tailings deposits within Dam I impoundment after stabilizing the dam illustrates that dam remediation is a difficult task that differs fundamentally from building a new dam. It akins to rehabilitate a patient. The eventual outcome depends on the health condition of the patient. Thus, the feasibility of such an endeavour as well as the selection of potential treatment methods requires careful evaluation by knowledgeable and experienced experts. Ordinary constraints such as time and budget have to defer to the safety-first priority. Potential consequences of any adopted treatment method have to be monitored closely to prevent unexpected incidents.

With the hindsight of Vale's unfortunate experience, it appears that the use of hydraulic dredges operating from the upstream shoreline of natural ground opposite to Dam I could be explored to economically dredge up tailings deposits for reprocessing. The operation would initiate far away from Dam I with minimum influence on the dam itself. The existing monitoring and alarm-warning system would be modified and improved so that vulnerable people in areas downstream of Dam I would be properly forewarned of any impending hazard to activate effective evacuation.

(4) Vale recently commissioned the Capanema Maximization Project, near Itabirito in Minas Gerais (see Fig. 3.142, Souza, 2024). The site is part of Vale's Operational Complex in Mariana and is expected to increase iron ore production

Fig. 3.142 Vale's Capanema Maximization Project, near Itabirito in Minas Gerais (from Souza, 2024)

by approximately 15 million metric tonnes per year towards the 2026 production guidance of 340–360 million Mt. Vale is to restart the Capanema project using dry processing and eliminating the use of tailings dams for tailings storage. Moreover, the project will produce sinter feed at low cost and will use autonomous trucks, which will improve safety and reduce emissions of greenhouse gas CO_2.

3.20 Case 19 March 28, 2020 Yichun Luming Tailings Spill, China

On March 28, 2020 at Yichun Luming Tailings Impoundment (see Figs. 3.143 and 3.144) in the Heilongjiang Province in northeast China, a large tailings spill occurred at 1:40 p.m. local time. The tailings release appeared to have originated from No. 4 Decant Tower near the north abutment of the south-segment of the Tailings Dam (see Fig. 3.144). The volume of spilled tailings was estimated at 2.5 million m^3.

Luming Tailings Impoundment Operation and Emergency Response

Luming Mine is operated by the China Railway Group. The mine started initial production in 2014, and had the daily process capacity of 45,000 tonnes of molybdenum ore. The tailings dam had a design height of 198 m, and reached 71 m height at the time of the tailings spill incident. The impoundment had a design storage capacity of 429 million m^3 tailings, and stored 64 million m^3 tailings in the spring of 2020. It was raised by the upstream construction method. The tailings impoundment had 10 decant towers over its design life for recycling supernatant water back to the beneficiating plant for mill use. In 2020, Decant Towers No. 1 and No. 2 were decommissioned at lower elevation, and Decant Towers No. 3 and No. 4 were in operation. Both No. 3 and No. 4 Decant Towers were equipped with video monitoring systems. However, the video monitoring system for No. 4 Decant Tower ceased to function on March 7, 2020 but was not repaired at the time of the incident.

Tailings spill from No. 4 Decant Tower system resulted in the tailings slurry plume entering into Yijimi River then Hulan River. The muddy colour of water downstream

Fig. 3.143 Satellite image of Luming Tailings Impoundment on March 25, 2020 prior to tailings spill (from Planet Labs, Petley, 2020)

Fig. 3.144 Satellite image of Luming Tailings Impoundment on March 29, 2020 after tailings spill (from Planet Labs, Petley, 2020)

of the tailings dam shown in Fig. 3.144 in contrast to the white ice cover in Fig. 3.143 indicated the sudden arrival of the emergency situation. Fortunately, there was no fatality.

The provincial government activated a 2nd-level environment emergency response at 9:00 a.m. on March 29, and took the following concurrent measures to control the spill and mitigate its downstream impact:

1. Stop No. 4 Decant Tower tailings leakage by infilling the void created by the collapsed tower—A 2 km long emergency road was constructed to improve the access of heavy equipment and construction materials to the worksite (see Figs. 3.145, 3.146 and 3.147). Two separate infilling operations were carried out at the collapsed tower on March 30 and March 31, using a plugged 0.6 m diameter, 2 m long steel pipe, cement, sand and stone, timber and geo-synthetic

fabric to infill the void. To ensure a robust repair, additional soil was added to build up an island over the infilled space followed by curtain grouting to complete the sealing operation on April 8.

2. Construct 18 cofferdams along Yijimi River and to intercept the escaped tailings slurry and to reduce its velocity and tailings concentration (see Fig. 3.148).
3. Monitor water quality of flow along Yijimi and Hulan River at various strategic locations downstream, and add flocculant and active carbon for water clarification (see Fig. 3.149). The target in molybdenum concentration was set at below 0.07 mg/L.
4. Replace polluted surface water sources of drinking water supply with unaffected groundwater sources in affected cities and upgrade No. 3 water treatment plant to replace the polluted No. 1 water treatment plant at Tieli city.

Fig. 3.145 Work site at the collapsed No. 4 Decant Tower (from https://www.thepaper.cn/newsDetail_forward_6772531)

Fig. 3.146 Tight construction site near No. 4 Decant Tower (from https://www.thepaper.cn/newsDetail_forward_6772531)

Fig. 3.147 Blocking No. 4 Decant Tower leakage (https://www.thepaper.cn/newsDetail_forward_6772531)

Fig. 3.148 Yimiji River polluted by spilled tailings slurry at a bridge site (https://finance.sina.com.cn/chanjing/gsnews/2020-04-03/doc-iimxyqwa4870999.shtml)

Fig. 3.149 Application of flocculant to river at a strategic location to reduce tailings concentration in water (https://finance.sina.com.cn/chanjing/gsnews/2020-04-03/doc-iimxyqwa4870999.shtml)

5. Stop polluted water in Hulan River entering Songhwa River 70 km upstream of the confluence on April 11, 2020. Thus, after two weeks of tireless efforts, the emergency response team of 9000 personnel (see Fig. 3.150) finally accomplished an important milestone i.e. prevented the polluted water entering Songhwa River. On May 2, 2020 Vice premier of Heilongjiang Province conducted a site review of the sealed Decant Tower No. 4 at the mine, and two water quality control points in affected downstream rivers (see Fig. 3.152). The total cost of the emergency management reached about 6.6 million US dollars, including 1.5 million US dollars of property loss.

Songhwa River is a major tributary of the border river Heilong River (Amur in Russia). Songhwa itself is also an important navigable stream in the province of Heilongjiang.

Failure Investigation Report

On May 20, 2021, about 14 months after the tailings spill incident the Joint Investigation Team consisted of the Heilongjiang Provincial Government, Ministry of Ecology and Environment and Ministry of Emergency Management issued its Investigation Report (2021). Key findings of the report are outlined and discussed in the following.

Causes of Tailings Spill

The main cause of the tailings spill is hinged on the quality of the design, operation and ongoing construction of No. 4 Decant Tower and its stop-logs. Figure 3.151 shows the reinforced concrete tower frame a day before the failure (Fig. 3.151a) and its close-up about 2 h after the discovery of the spill (Fig. 3.151b). Photos of fragments of stop-logs collected after the spill are shown in Fig. 3.151c, d. Details of stop-log dimensions and reinforcement steels, as shown on Fig. 3.153, indicate serious discrepancy between the design and as-built.

No. 4 Decant Tower reinforced concrete frame was 21 m tall, and consisted of 6 vertical columns and 7-level ring beams spaced at 3 m intervals. At the time of incident, about 300 stop-logs were installed in the tower, and the tailings level reached

Fig. 3.150 Emergency response headquarter coordinated all efforts of a large project team (from http://www.jxbbs.cc/wap/37114-1.html)

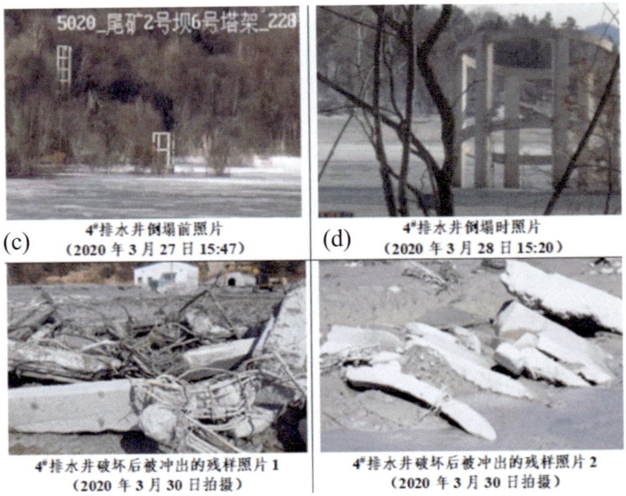

(c) (d)

(a) No. 4 Decant Tower before tailings spill
(b) No. 4 Decant Tower during tailings spill
(c) Damaged reinforced concrete stop-logs flushed out from the damaged decant system
(d) Flushed out fragments of stop-logs

Fig. 3.151 Critical photos showing deficiencies of the reinforced concrete No. 4 Decant Tower and stop logs (from Luming Investigation Report, 2021). **a** No. 4 Decant Tower before tailings spill. **b** No. 4 Decant Tower during tailings spill. **c** Damaged reinforced concrete stop-logs flushed out from the damaged decant system. **d** Flushed out fragments of stop-logs

Fig. 3.152 Site review by Vice Premier Li of Heilongjiang Province on May 2, 2020 (https://www.thepaper.cn/newsDetail_forward_7447992)

13.63 m height. Discrepancy between the design and as-built ring-beam height as well as deficiencies of reinforcement steels of both the columns and ring beams and poor workmanship are built-in weakness of the tower frame. As tailings started spilling through the tower due to structural failure of stop-logs, the decant tower

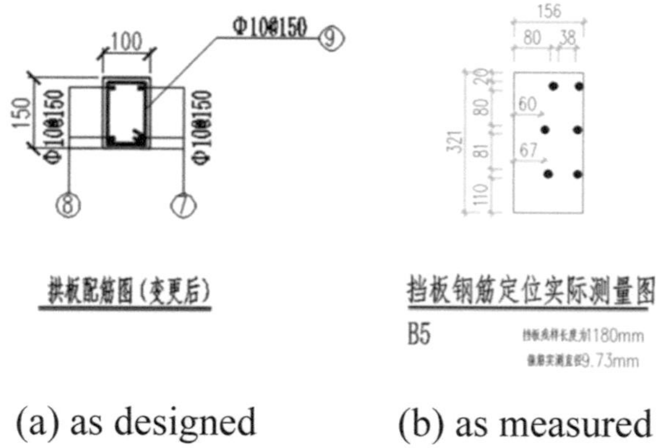

Fig. 3.153 Dimensions and reinforcement steels of concrete stop-log used to seal decant tower segments below tailings level (from Luming Investigation Report, 2021). **a** As designed. **b** As measured

began to tilt due to the unbalanced forces from water, tailings and ice cover, and collapsed soon after.

Systematic Issues

The investigation also revealed pervasive systematic issues that influenced the performance of the tailings impoundment supernatant water recovery system.

It appeared that the mine outsourced its decant towers design, construction and quality control to three contractors whose qualification as well as actual performance had not been scrutinized. Some of the deficiencies discovered in No. 4 Decant Tower were also observed in earlier years at decommissioned No. 1 and No. 2 Decant Towers.

Overall, 52 persons within the mine and its contractors were held responsible for the tailings spill incident. In addition, 11 persons in the government regulatory agencies were also reprimanded for not carrying out their monitoring and overseeing responsibilities (Sixth Tone, 2021). Luming Mine was fined 13.26 million yuan (equivalent of US $2.1 million) for causing this tailings spill environmental incident (Mining.com, 2021).

Lessons Learned

This incident highlighted importance of reclaim systems of supernatant water to the safety and integrity of a tailings impoundment. Fortunately, with massive emergency measures taken promptly, the collapsed decant tower was plugged to prevent further tailings loss or even breach of the tailings dam. The eventual accomplishment of stopping spilled tailings from entering Songhwa River, a tributary of the Heilong River (a border river between China and Russia) did contain the river pollution problem within the province. While the emergency response was a hard-earned success, the reason for the creation of such an emergency was an inexcusable example of tailings management failure. It is hoped that the mine would be jolted by this major incident, and would strive to tighten up all aspects of its tailings operation under more stringent oversight by regulatory agencies.

This case history illustrates the importance of the integrity of the supernatant water management system. The assurance of a safe and robust water management system is crucial to safeguard the areas downstream of the tailings impoundment free from inundation and environmental degradation.

3.21 Case 20 Heavy Leakage of Piney Point Industrial Wastewater Storage Pond, USA

The Governor of Florida, Mr. Ron DeSantis declared a state of emergency on April 3, 2021, after a significant leak at a large pond of wastewater that threatened to flood roads and homes in the Manatee County (see AP, 2021 and Figs. 3.154 and 3.155). The location of Piney Point Industrial Wastewater Storage Ponds near Tempa Bay is shown in Fig. 3.156, and the satellite image of the wastewater holding ponds in October, 2020 is given in Fig. 3.157. The emergency evacuation order closed off a highway and involved more than 300 homes near the leaky pond in the Tampa Bay area north of Bradenton. The state indicated a leak was detected in the previous day involving wastewater containing phosphorus and nitrogen from the old Piney Point phosphate plant (see Fig. 3.157). Pumps were brought in to drain the leaky pond, which contained 480 million gallons of seawater mixed with legacy process water and stormwater runoff (see Figs. 3.158 and 3.159).

The evacuation order was lifted in the afternoon of April 6, as the emergency pumping had removed sufficient volume of water from the leaky pond and averted

Fig. 3.154 Governor DeSantis, state and local officials at a news conference on Apr 4 at the Manatee County Emergency Management Office in Palmetto, Florida (from NPR, 2021)

Fig. 3.155 Leaky New Gypsum Stack South Pond (from Aklib, 2021)

Fig. 3.156 Location of Piney Point phosphate processing facility near Tampa Bay, Florida (from Bausback, 2022)

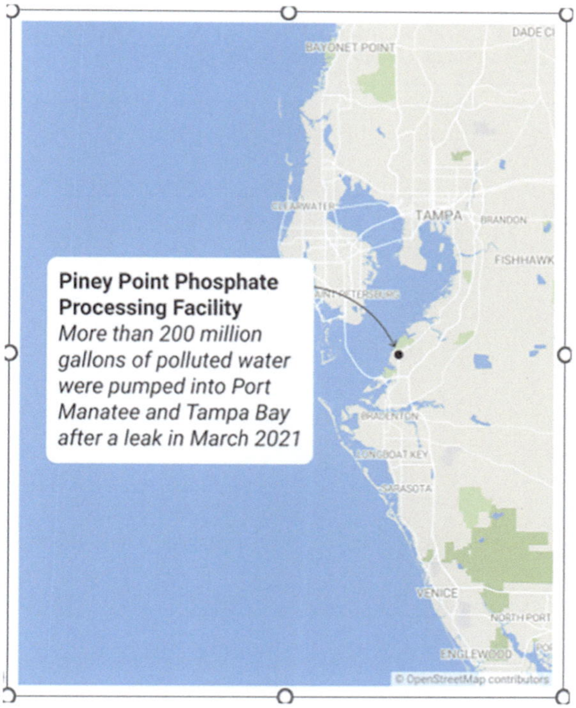

Fig. 3.157 Satellite image of Piney Point wastewater holding ponds in October, 2020 (Maxar Tech/AFP/Getty Images, from Guardian, 2021)

Fig. 3.158 Storage ponds on Apr 2, 2021, one day before the issuance of an evacuation order (from AP, 2021, Apr 3)

Fig. 3.159 Emergency pumping water from the leaky New Gypsum Stack South Pond reduced the pond volume to less than 300 million gallons (from Bradenton Herald, 2021)

the immediate crisis of potential collapse of the radioactive phosphogypsum stacks beneath the pond (see Fig. 3.160).

Phosphogypsum Stack and Industrial Wastewater Storage Ponds

Piney Point phosphogypsum stack is a storage facility of solid chemical waste produced in the process of fertilizer from phosphate rock. Phosphogypsum is weakly radioactive due to the presence of uranium and thorium, and their daughter isotopes radium, radon, polonium, etc. Figure 3.160 is a schematic sketch of a typical wastewater storage pond atop of radioactive phosphogypsum stack. Phosphate production is

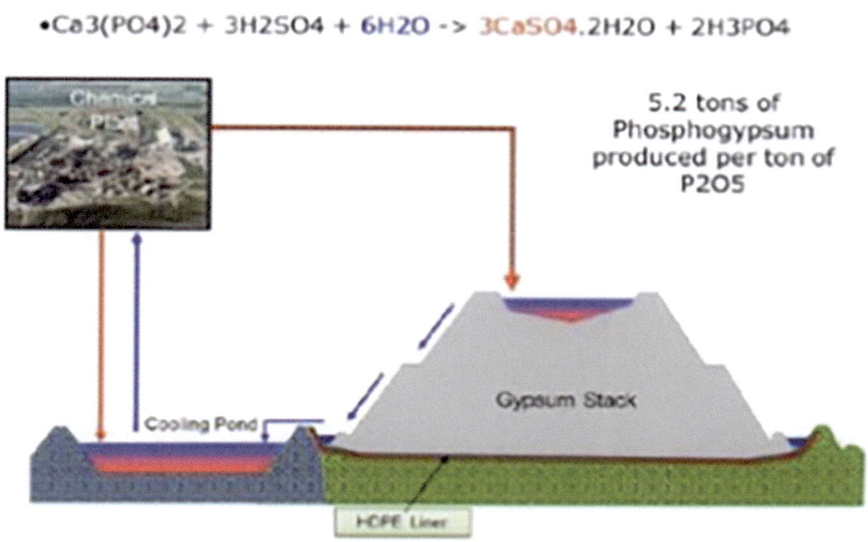

Fig. 3.160 Schematic sketch of wastewater storage pond atop of radioactive phosphogypsum stack (https://nucleus.iaea.org/sites/orpnet/home/Shared%20Documents/T1-Beckingham-Phosphogypsum-Production-USA.pdf)

a major industry in Florida, thus phosphogypsum stacks are also relatively common in the state.

Dailykos (2021) provided a detailed description of the troubled history associated with the Piney Point phosphogypsum stack and wastewater storage ponds. Because of financial difficulties, the facility has not been properly closed. It has experienced changes of ownership, bankruptcies, leaks of pond water, repairs of pond liner, and emergency dumping of pond water to the adjacent Tampa Bay. Figure 3.161 shows some of these activities in 2004–2005.

The 2021 heavy leakage from the Piney Point wastewater storage ponds atop of phosphogypsum stack threatened to flood road and homes in the surrounding area was, in fact, the latest incident occurred at this facility. The incident resulted in the declaration of an evacuation order and emergency dumping of pond water to the nearby Tempa Bay.

Bausback (2022) provided a multi-decade timeline of the saga of Piney Point Industrial Wastewater environmental issue which began with the opening of a fertilizer process plant by the Borden Chemical Company in September 1966. Over the last six decades or so, the issue plagued the local community and the Florida state for the following two reasons:

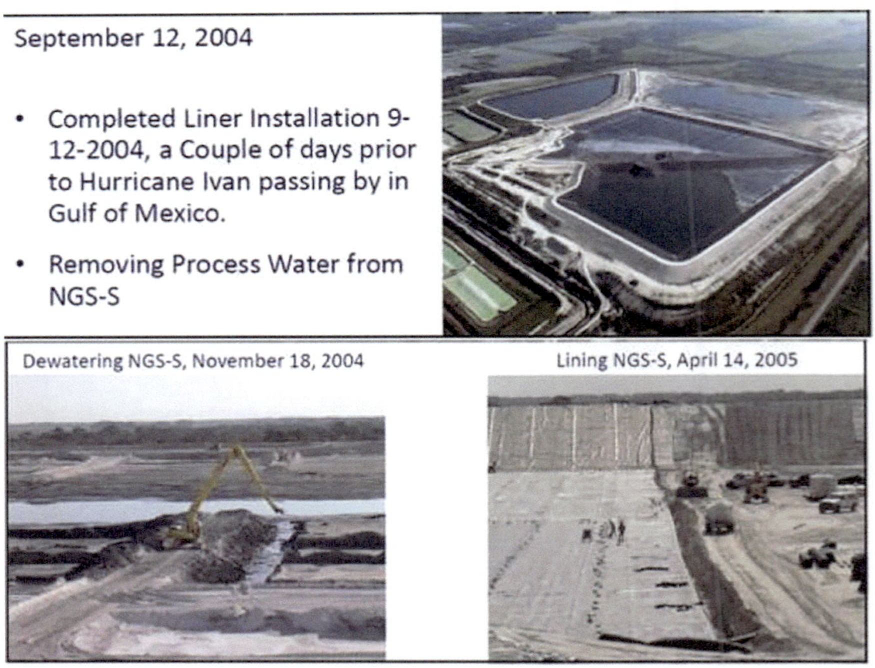

Fig. 3.161 2004–2005 construction work at wastewater storage pond related to Hurricane Ivan (https://www.tbrpc.org/wp-content/uploads/2020/12/2020-12-10_PineyPoint_Update.pdf)

1. The facility was not properly closed, and has become the source of perennial environmental problem; and
2. The facility went through many ownership changes and bankruptcies, and its stewardship has been handicapped by inadequate funding.

The 2021 leakage incident was successfully handled by emergency evacuation, pond dewatering and release of excess pond water to Tempa Bay with the involvement of local and state governments.

Developments After 2021 Emergency Discharge of Excess Pond Water to Tempa Bay

The 2021 emergency discharge of excess pond water to Tempa Bay appeared to have caused algal blooms including a massive "red tide" fish kill. Since then, the following two significant developments about the industrial wastewater environmental issue are worthy of note.

1. Lake (2023) reported the closure of the first of 4 wastewater storage ponds. Figure 3.162 shows the bird's-eye view of the closed pond. A liner has been installed on top of the closed pond. A new stormwater control system has been installed and grass has been planted to reduce soil erosion. A scheme of transfer pond water to an underground aquifer after removal of its nitrogen and phosphorus content is an essential component of this closure strategy. On Dec 16, 2021, a permit to Manatee County to build an underground injection well on Buckeye Road to receive industrial wastewater from Piney Point. The well began receiving the wastewater from the site on April 4, 2023. By October 2023, more than 97 million gallons of wastewater have been transferred to the well.
2. Ballogg (2024) reported the settlement of a federal lawsuit between environmental groups and local and state governments of Florida over pollution related to the 2021 emergency discharge of pond water to Tempa Bay. The settlement agreement includes: active implementation of closures of all 4 wastewater storage ponds, promulgation of a Clean Water Act permit to monitor and report publicly on potential sources of pollution at Piney Point. Based on the progress of current closure activities, the completion of closures of all 4 storage ponds is estimated in 2025.

Lessons Learned

The 2021 Piney Point emergency evacuation and discharge of industrial wastewater to Tampa Bay illustrate the following key lessons:

1. Water storage in a tailings facility (phosphogypsum stack in this case) is often a major contributor to safety issue of the facility. The fact that the stored water contains nitrogen and phosphorus nutrient and potentially radioactive isotopes make the issue more severe.
2. The original phosphate fertilizer process plant at Piney Point was not closed properly. The ownership changes and bankruptcies and lack of funding further handicapped care and maintenance of the facility.

Fig. 3.162 Closure of the first of four Pitney Point industrial wastewater storage ponds (from Lake, 2023)

3. The closure of the first storage pond in 2023 and ongoing work aiming at the closure of remaining 3 storage ponds in 2025 bring hopes to a satisfactory ending of the multi-decade saga of industrial wastewater environmental issue.
4. This case could serve as an example of solving phosphogypsum stack environmental issue in other Florida localities.

3.22 Case 21 Mine Life of Żelazny Most Tailings Storage Facilities, Poland

The Żelazny Most Tailings Impoundment is located in south-west of Poland surrounded by woodland and arable land (see Figs. 3.163, 3.164 and 3.165). It has been in operation by KGHM Polska Miedź, a state owned mining and milling company for extracting copper since 1977. The reason to include this case here is not to cover a dam failure. Rather, it is to show a success story of how this tailings dam survived over 47 years to date, overcoming various difficulties including: a complex geological setting, potential liquefaction of tailings dam and substantial economic transition of geopolitical setting in the 1990s among others.

Fig. 3.163 Location of the Żelazny Most copper mine (from Jamiolkowski, 2014)

Fig. 3.164 Aerial view of the Żelazny Most tailings impoundment (from Jamiolkowski, 2014)

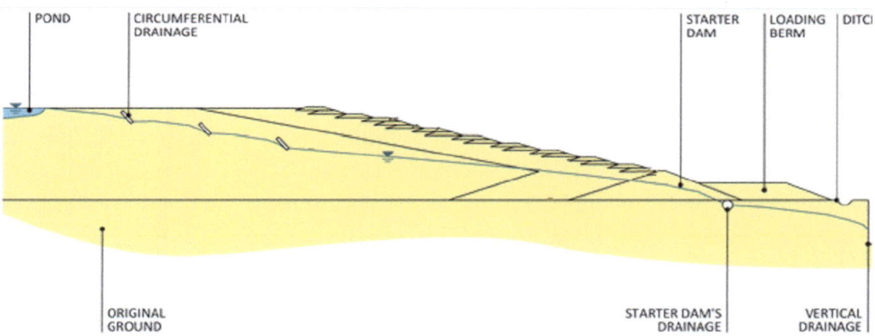

Fig. 3.165 Typical cross-section of the Żelazny Most tailings dam (from Zieliski & Stefanek, 2021)

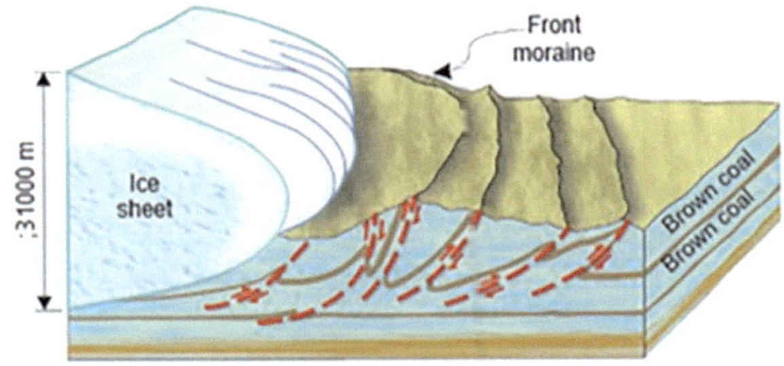

Fig. 3.166 Historic foundation deformation ahead of a moving ice sheet (from Jamiolkowski, 2014)

Żelazny Most Tailings Storage Facility

The tailings impoundment is enclosed by a ring-shaped tailings dam having a perimeter of 14 km and footprint area of 16 km^2 (Stefanek et al. 2017). The foundation subsurface profile in the impoundment area is from the ground surface downward following a geological sequence as follows (Fig. 3.166):

- Superficial solifluction deposits due to freeze-thaw process over local areas;
- Quaternary deposits including Pleistocene glacial outwash sand and gravel, and stiff clays; and
- Neogene deposits involving generally stiff clays.

The highly plastic Neogene stiff clays had been significantly influenced by the northwest-towards-southeast Quaternary glacier movements resulting extensive development of slickensided shear surfaces with its drained shear strength reduced to much lower residual value (see Fig. 3.170).

The Żelazny Most tailings dam design and construction follows Poland engineering practice and the observational method promulgated by Terzaghi and Peck (Peck, 1969b). In 1992, on the advice of the World Bank, KGHM appointed an international board of experts chaired by late Professor Jamiolkowski to oversee the safe development of the tailings dam. The tailings dam was constructed by the upstream dam method with 5-levels of circumferential internal drains (see Figs. 3.167 and 3.168). The dam crest has been raised at a rate of about 1.0–1.5 m/year, with an average downstream slope of 3.5 horizontal to 1 vertical. Toe buttress berms as well as additional slope-flattening berms and a row of relief wells were added as required to stabilize the ring dam during its construction.

The minimum upstream tailings beach width, i.e., the distance from the dam crest to the tailings pond edge, is maintained at 200 m. As shown in Fig. 3.167, the deposited tailings material adjacent to the Starter Dam or Confinement Tailings Dykes is fine to medium sand. Moving towards the upstream decant pond, the deposited material grades finer into silty sand then sandy silt. Figure 3.169 shows the tailings impoundment in plan view, while Fig. 3.170 presents a simplified cross section of the East Tailings dam and foundation. Extensive piezometers are installed in the impoundment (see Fig. 3.171) for monitoring pore-pressure conditions in the dam and its foundation. Pressure-relief wells and drainage wells used to control seepage

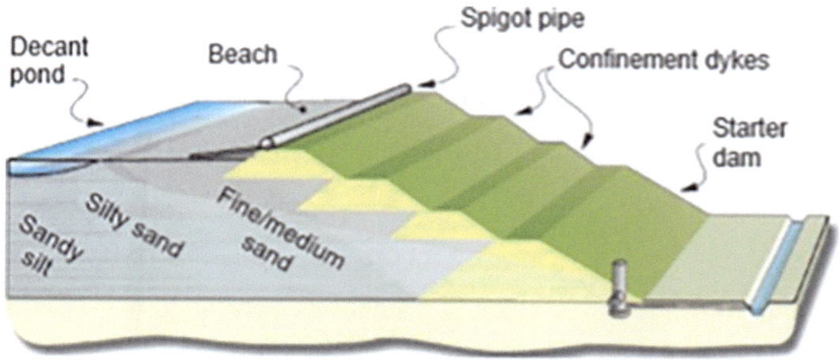

Fig. 3.167 Schematic cross section of tailings dam and tailings impoundment (from Jamiolkowski, 2014)

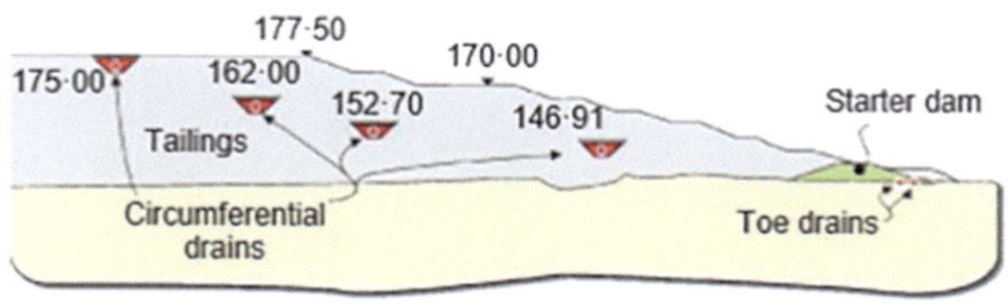

Fig. 3.168 Level circumferential internal drains (from Jamiolkowski, 2014)

pressure as well as to reduce seepage bypass to the downstream surrounding areas are shown in Fig. 3.172.

Geotechnical Studies

In his prestigious 53rd Rankine Lecture, late Professor Jamiolkowski (2014) summarized considerable studies on the foundation soils and tailings damfill of the impoundment, including both laboratory and field insitu testings, and finite-element dam stability and deformation analyses.

Figure 3.173 shows results of a series of anisotropically consolidated undrained compression triaxial shear tests on undisturbed tailings samples retrieved from the tailings impoundment with the axial consolidation stress in the range of 50–1000 kPa (Lipinski, 2012). The figure indicates that the tailings undrained shear behaviour changes from dilative to contractive-dilative or even contractive as the consolidation stress increases from 50 to 1000 kPa. Moreover, as shown in Figs. 3.167 and 3.170, as the height of dam is being raised, the bulk of tailings dam involves more fine-grained and weaker tailings material. Thus, even tailings had not been a major focus of attention so far at Żelazny Most, its susceptibility to static and/or seismic liquefaction is expected to play a more important role in the future performance of the impoundment whose peripheral ring-shaped tailings dam had been constructed by the upstream dam method.

Under the guidance of the International Board of Experts, the ongoing operation of the tailings impoundment and dam raising activities have been underpinned by a thorough dam monitoring program and ongoing confirmation dam stability and

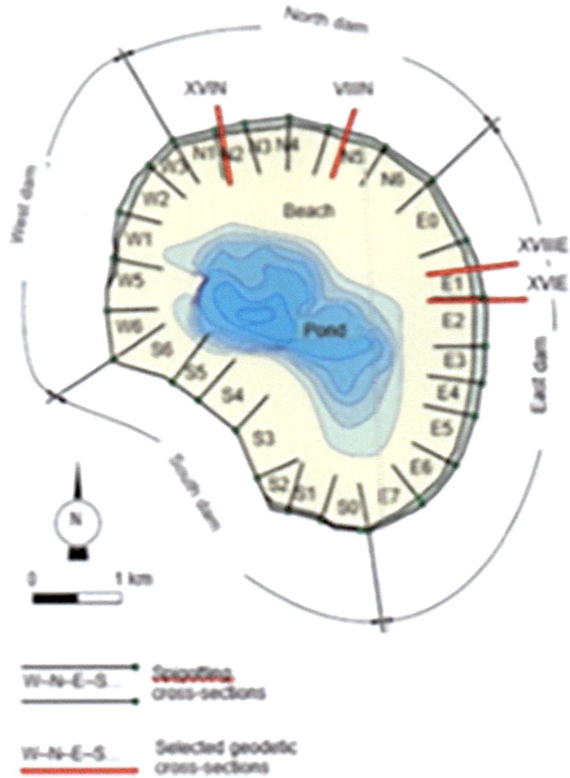

Fig. 3.169 Plan view of tailings impoundment (from Jamiolkowski, 2014)

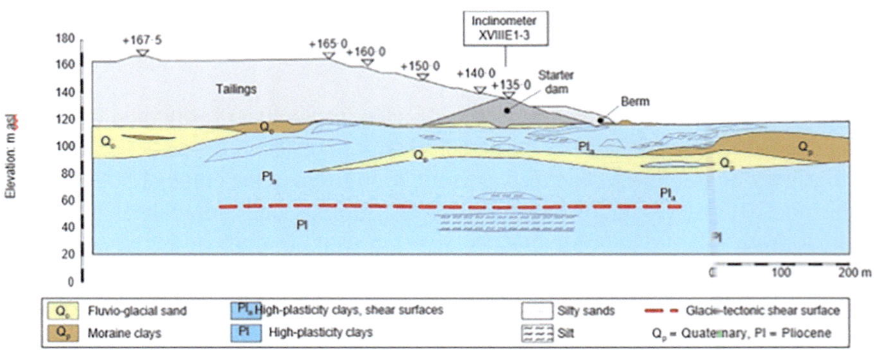

Fig. 3.170 East tailings dam—simplified dam section and subsoil profile (from Jamiolkowski, 2014)

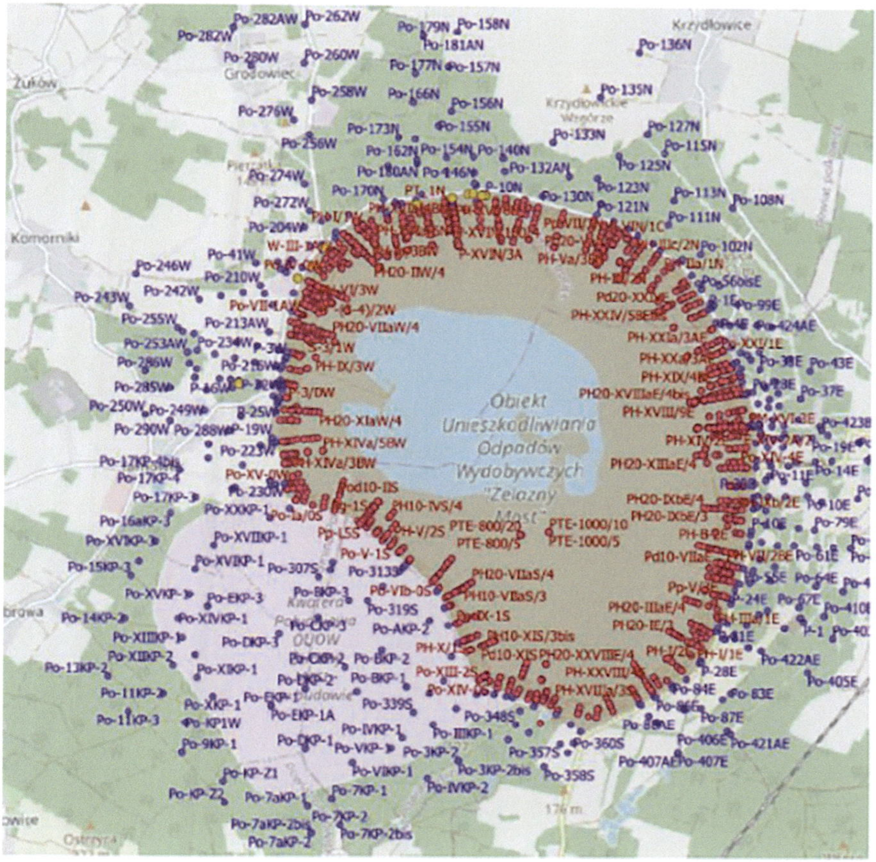

Fig. 3.171 Locations of piezometers for monitoring pore pressure in the Żelazny Most tailings impoundment (from Zieliski & Stefanek, 2021)

deformation analyses. The dam surveillance and monitoring program consists of the following:

- Water level of the supernatant pond, pore water pressures in the damfill and foundation;
- Displacements of the dam and foundation;
- Deep and shallow inclinometers to measure horizontal displacement profiles;
- Discharges of drainages;
- Seismic vibrations;
- Visual assessment of the tailings dam, tailings beach, and downstream area; and
- Impoundment water balance.

Dam analyses have been carried out using soil strength parameters and stress–strain relationships based on laboratory and field insitu testing results. The East and North tailings dams have been modified by flattening the overall downstream slope

Fig. 3.172 Locations of pressure-relief and drainage wells for Żelazny Most tailings impoundment (from Zieliski & Stefanek, 2021)

and using toe-buttress berms. In addition, dam raising activities have been suspended for 3–5 years starting from 2017 in order to confirm that pore water pressures induced by dam construction in the foundation soils would have adequate time to dissipate especially at the North Dam.

Future Tailings Storage

A recent satellite imagery of Żelazny Most Tailings Impoundments is shown on Fig. 3.174. Stefanek et al. (2017) provided an overview on options for accommodating additional tailings storage volume in the future for a life-of-mine plan to 2050. Considered options include the following:

- Raise the existing dam crest above the present elevation of ~ 180 m;
- Extend the existing impoundment towards south as the Southern Extension; and
- Add an additional tailings impoundment at Gilow site.

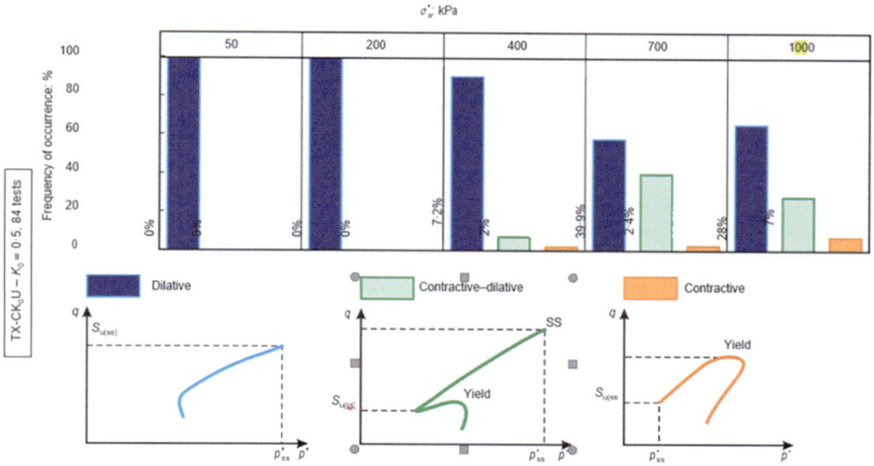

Fig. 3.173 Tailings undrained triaxial shear test results (from Jamiolkowski, 2014)

Fig. 3.174 Recent satellite imagery of Żelazny Most tailings impoundments (from Google Earth Pro)

Different construction method other than the upstream dam method will be considered for the new impoundment at the Southern Extension or Gilow site, including using thickened tailings.

Lessons Learned

The success story of the Żelazny Most tailings impoundment is by no means accidental. Fortunately, the two major issues that threaten its safety: i.e., potential liquefaction of tailings damfill and low shear strength of highly plastic foundation clay were recognized early and dealt with satisfactorily up to now. Under the guidance of the International Board of Experts led by the renowned late Professor Jamiolkowski, the stability of the ring-shaped tailings dam has been maintained by a thorough dam monitoring program and confirmation dam stability analyses using up-to-date monitored data. Measures to reduce the environmental impact of the Żelazny Most Tailings Storage Facility to the mine site and its surrounding areas are highlighted in Zieliski and Stefanek (2021).

The current success enjoyed by the Żelazny Most tailings storage facility, however, needs redoubled effort in order to complete the project for the entire life-of-mine to 2050. For future construction in the Southern Extension or Gilow site, centreline construction method using cycloned tailings sand as damfill could be both economical and flexible to meet further challenges of safe dam-raising program. The recent dam failures since 2014 include: foundation failure in Case 15, static liquefaction flow failures in Cases 16, 17 and 18. All these cases point to the need of adopting the safer centreline construction method (Lo et al. 2003). The sooner the stoppage of the upstream dam construction method at the Żelazny Most, the better its future safety prospects. Above all, the leadership of the International Board of Experts is to be rigorously followed.

References

AECOM. (2009, June 25). Root cause analysis of TVA Kingston dredge pond failure on Dec. 22, 2008.

AP. (2021, April 3). *Leak at wastewater pond prompts evacuations in Florida*. Associated Press.

Baker, F., Cook, D., Deem, J., Letient, H., & Walsh, H. (1996). Matachewan mines tailings dam—From failure to restoration. In *Proceedings of CDSA/CANCOLD Conference*.

Ballogg, R. (2024, June 9). Florida agrees to settlement in Piney Point pollution lawsuit. Here's what it says. *Miami Herald*.

Banks, G. A. (2001, October). *Papua New Guinea baseline study*. University of New South Wales, Australia, https://www.researchgate.net/publication/265553794_Papua_New_Guinea_Baseline_Study

Bausback, E. (2022, April 22). *A timeline of the Piney Point wastewater disaster*. University of Florida Thompson Earth Systems Institute.

Berti, G., et al. (1988). The disaster of Stava/Northern Italy, hydraulic fill structures. *ASCE Geotechnical Special Publications, 21*, 492–510.

Blight, G. E., & Fourie, A. B. (2003). *A review of catastrophic flow failures of deposits of mine waste and municipal refuse*. http://ww.unina2.it/flows2003/flows2003/A.B.%20FOURIE.pdf

Boroschek, R. L., Contreras, V., Kwak D. Y., & Stewart, J. P. (2012). Strong ground motion attributes of the 2010 Mw 8.8 Maule, Chile, earthquake. *Earthquake Spectra, 28*(S1), S19–S38.

Bradenton Herald. (2021, April 5/6). Will a deep well put an end to Piney Point? State officials say funds are available.

Brumadinho Engineering History. (2020). *World mine tailings failures—From 1915*. Retrieved September 23, 2023, from https://worldminetailingsfailures.org/corrego-do-feijao-tailings-failure-1-25-2019/

Burke, H. H., & Smucha, S. S. (1979). Lornex tailings dam on a soft foundation. In *Proceedings of 6th Panamerican Conference on Soil Mechanics and Foundation Engineering* (pp. 305–316).

Byrne, P., & Seid-Karbasi, M. (2003). Seismic stability of impoundments. In *17th Annual Symposium*. Vancouver Geotechnical Society.

Carrera, A., et al. (2011). Influence of grading on the mechanical behaviour of Stava tailings. *Geotechnique, 61*(11), 935–946.

Chandler, R. J., & Tosatti, G. (1995). The Stava tailings dams failure, Italy, July 1985. *Proceedings of the Institution of Civil Engineers—Geotechnical Engineering, 113*(2), 67–79.

Chua, S. M. T., et al. (2019, October). Leveraging open-access remote sensing imagery to monitor dam infrastructure: Case study of the Cadia tailings dam collapse, Australia. In *NZSOLD/ANCOLD 2019 Combined Conference*, Auckland.

CIMNE. (2021). *Computational analyses of Dam I failure at the Corrego de Feijao mine in Brumadinho* (Final report, with Appendices, 561 pp.). International Centre for Numerical Methods in Engineering.

Cohen, M. W., & Moenne, G. B. (1992). Tailings deposits at the El Soldado Mine Compania Minera Disputada de las Condes. In *Seismic design, abandonment and rehabilitation of tailings dams* (Special Volume, pp. 199–242). Sociedad Chilena de Geotecnia, Vina del Mar, Chile.

Colombo, P., & Colleselli, F. (2003). The collapse of the Stava tailings basins. Analysis of the event and observations. In G. Tosatti (Ed.), *A review of scientific contributions on the Stava Valley disaster (Eastern Italian Alps), 19th July 1985, Pitagora, Bologna* (pp. 335–358).

Dailykos. (2021, April 5). Updated: Phosphogypsum stacks, toxic ponds, leaks and spills, sadly, Manatee County has it all. https://www.dailykos.com/stories/2021/4/5/2024366/-Phosphogypsum-stacks-toxic-ponds-leaks-and-spills-unfortunately-Manatee-County-has-it-all

Das, S., Priyadarshana, A., & Grebby, S. (2024). Monitoring the risk of a tailings dam collapse through spectral analysis of satellite InSAR time-series data. *Stochastic Environmental Research and Risk Assessment, 38*, 2911–2926. https://doi.org/10.1007/s00477-024-02713-3

Davies, W. E. (1973). Buffalo Creek dam disaster: Why it happened. *Civil Engineering, 43*(7), 69–72.

Dobry, R., & Alvarez, L. (1967). Seismic failures of Chilean tailings dams. *Journal of the Soil Mechanics and Foundations Division, ASCE, 93*(6), 237–260.

Elkhamra, Y., Chen, H., & Stark, T. (2023). Inverse analysis of Cadia tailings dam failure. In *Geo-Congress 2023*. https://doi.org/10.1061/9780784484708.002

ENR. (2014, October 6). *B.C. tailings dam spillage bigger than initial estimates*. Engineering News Record.

ENR. (2021a, April 6). Environmental concerns loom as leaking Tampa Bay Reservoir evacuation order lifts. *Engineering-News Record*.

ENR. (2021b, April 13). UPDATE: Fla. Governor closes leaking phosphate waste site, $15.4M for emergency cleanup. *Engineering-News Record*.

EPA. (2014, December). 2014 final rule: Disposal of coal combustion residuals from electric utilities. https://www.epa.gov/sites/default/files/2015-03/documents/ccr_webinar_slides_508.pdf

Expert Panel. (2019, December 12). *Report on the technical causes of the failure of Feijão Dastrum I*. http://www.b1technicalinvestigation.com/

Fourie, A. B., Blight, G. E., & Papageorgiou, G. (2001). Static liquefaction as a possible explanation for the Merriespruit tailings dam failure. *Canadian Geotechnical Journal, 38*(4), 707–719.

Fundão Tailings Dam Review Panel. (2016, August 25). *Report on the immediate causes of the failure of the Fundão Dam*. https://www.resolutionmineeis.us/docuents/fundao-2016

Gebhart, T. R. (2016, September). *Post-liquefaction residual strength assessment of the Las Palmas, Chile tailing failure* [Master thesis]. California Polytechnic State University.

GEER. (2010, March 25). *Reconnaissance of the Feb 27, 2010 Maule, Chile, earthquake, version 2*.

Griffiths, J. S., Hutchinson, J. N., Brunsden, D., Petley, D. J., & Fookes, P. G. (2004). The reactivation of a landslide during the construction of the Ok Ma tailings dam, Papua New Guinea. *Quarterly Journal of Engineering Geology and Hydrogeology, 37*, 173–186.

Guardian. (2021, April 11). No community should suffer like this: Florida's toxic breach was decades in the making.

Hambrett, M. (2021). Newcrest's Cadia gold mine tailings dam collapse in 2018 causing concern. *ABC Central West*.

Hettler, J., Irion, G., & Lehmann, B. (1997). Environmental impact of mining waste disposal on a tropical lowland river system: A case study on the Ok Tedi Mine, Papua New Guinea. *Mineralium Deposita, 32*, 280–291.

Holden, D., Donegan, S., & Pon, A. (2020). Brumadinho Dam InSAR study: Analysis of TerraSARX, COSMO-SkyMed and Sentinel-1 images preceding the collapse. In P. M. Dight (Ed.), *Proceedings 2020 International Symposium on Slope Stability in Open Pit Mining and Civil Engineering* (pp. 293–306). Australian Centre for Geomechanics. https://papers.acg.uwa.edu.au/p/2025_15_Holden/

Hosgood, A. F. (2024, December 10). A decade later, imperial metals faces consequences for the Mount Polley disaster. *The Tyee*. https://thetyee.ca/News/2024/12/10/Mount-Polley-Disaster-Consequences/

Ishihara, K. (1984, May 6–11). Post-earthquake failure of a tailings dam due to liquefaction of the pond deposit. In *Proceedings of International Conference on Case Histories in Geotechnical Engineering*, Rolla, Missouri (Vol. 3, pp. 1129–1143).

ITRB. (2019, April 17). *Report on northern tailings storage facility embankment failure—Cadia Valley Operations*.

Jamiolkowski, M. (2014). Soil mechanics and the observational method: Challenges at the Zelazny Most copper tailings disposal facility. *Geotechnique, 64*(8), 590–618.

Jodah, D. K. (1995). Courting disaster in Guayana. *Multinational Monitoring, 16*(11).

Klohn, E. J. (1979). Seepage control for tailings dams. In *Proceedings of 1st International Mine Drainage Symposium*, Denver, Colorado.

Klohn, E. J., & Lo, R. C. (1978). Tailings dams on plastic clay foundations. In *Proceedings of 31st Canadian Geotechnical Conference*, Winnipeg, Manitoba (pp. 4.1.1–4.1.26).

Klohn, E. J., Lo, R. C., & Olsen, M. T. (1982, May). Stability of a tailings dam on sensitive clay. In *Transactions of the 14th International Congress on Large Dams* (Vol. II, pp. 199–217), Rio de Janeiro.

Klohn, E. J., Portfors, E. A., & Lo, R. C. (1987, May). Siting a major tailings dam in difficult mountainous terrain. In *International Symposium on Engineering Geology and the Environment in Mountainous Areas*, Beijing, China (Vol. 1, pp. 293–308).

Lake, L. (2023, October 18). First wastewater pond closed at Piney Point. https://www.amisun.com/2023/10/18/first-wastewater-pond-closed-at-piney-point/

Liber Media Brasil. (2019, February 1). Momento exato do rompimento da barragem da Vale, a Portuguese broadcast. https://www.youtube.com/watch?v=SZ8ma4XOgow

Lipinski, M. J. (2012). *Laboratory tests on undisturbed samples of tailings from Zelazny Most tailings pond* (Report 144/2645/10). Geoteko Geotechnical Consultants Ltd., Warsaw, Poland.

Lo, R. C., McLeod, H. N., & Klohn, E. J. (2003). Centreline tailings dam—Overview of a successful design practice. In *Proceedings of ICOLD Symposium on Major Challenges in Tailings Dams*, Montreal.

Luino, F., & De Graff, J. (2012). The Stava mudflow of 19 July 1985 (Northern Italy): A disaster that effective regulation might have prevented. *Natural Hazards and Earth System Sciences, 12*(4), 1029–1044. Stava Foundation Archive 1985. www.stava1985.it

Luming (Mine Tailings Spill) Investigation Report 2021 黑龙江伊春鹿鸣矿业有限公司"3·28"尾矿库泄漏次生重大突发环境事件调查报告 2021, May 20 (8 pp.). http://mee.gov.cn/ywgz/hjyj/yjxy/202105/t20210520_833709.shtml

Macedo, J., Yerro, A., Renzo, C., & Pierce, I. (2024). Cadia TSF failure assessment considering triggering and posttriggering mechanisms. https://doi.org/10.1061/JGGEFK.GTENG-11

Marcuson, W. F., Ballard, R. F., & Ledbetter, R. H. (1979). Liquefaction failure of tailings dams resulting from the near Izu Ohshima earthquake, 14 and 15 January, 1978. In *Proceedings of 6th Panamerican Conference on Soil Mechanics and Foundation Engineering*, Lima (Vol. 2, pp. 69–80).
McDermott, R. K., & Sibley, J. M. (2000). The Aznalcóllar tailings dam accident—A case study. *Mineral Resources Engineering, 9*(1), 101–118.
Michael, P. R., et al. (2000). Potential of breakthroughs of impounded coal refuse slurry into underground mines. *Environmental & Engineering Geoscience, XVI*(3), 299–314.
Minerals Councils. (2020). *We care and we remember Merriespruit 22 February 11, 2014*. South Africa Minerals Council.
Mining.com. (2021, August 17). Chinese miner fined over molybdenum ore spill—Media. https://www.mining.com/web/chinese-miner-fined-over-molybdenum-ore-spill-media/
MSHA. (2003). *Guidance for evaluating the potential for breakthroughs from impoundments into underground mine workings and breakthrough prevention measures, Appendix D*. Mine Safety and Health Administration (MSHA).
MSHA. (2009). *Engineering and design manual—Coal refuse disposal facilities* (2nd Ed.). D'Appolonia Consulting Engineers.
Mudd, G. M., Roche, C., Northey, S. A., Jowitt, S. M., & Gamato, G. (2020). Mining in Papua New Guinea: A complex story of trends, impacts and governance. *Science of the Total Environment, 741*.
Murray, L., Thompson, M., & Lane, M. (2003, June 15). Obstacles to on-land mine tailings disposal in Papua New Guinea. In *ICOLD 2003, International Symposium on Major Challenges in Tailings Dams*, Montreal.
NPR. (2021, April 4). State of emergency, evacuations in Florida County over wastewater pond leak.
NRC. (2002). *Coal waste impoundments—Risks*. National Academy Press.
NY Times. (2019, February 9). A tidal wave of mud. https://www.nytimes.com/interactive/2019/02/09/world/americas/brazil-dam-collapse.html
Okusa, S., & Anma, S. (1980). Slope failures and tailings dam damage in the 1978 Izu-Ohshima-Kinkai earthquake. *Journal of Engineering Geology, 16*, 195–224.
Okusa, S., Anma, S., & Maikuma, M. (1984, July 21–28). The propagation of liquefaction pressure and delayed failure of a tailings dam dike in the 1978 Izu-Oshima-Kinkai earthquake. In *Proceedings of 8th World Conference on Earthquake Engineering*, San Francisco (Vol. 1, pp. 389–396).
Olalla, C., & Cuellar, V. (2001). Failure mechanism of the Aznalcóllar Dam, Seville, Spain. *Geotechnique, 51*(5), 399–406.
Peck, R. B. (1969a). Advantages and limitations of the observational method in applied soil mechanics, 9th Rankine lecture. *Geotechnique, 19*(2), 171–187.
Peck, R. B. (1969b). Advantages and limitations of the observational method in applied soil mechanics. *Geotechnique, 19*(1), 171–187.
PEER. (2019, January). *Flow-failure case history of the Las Palmas, Chile, tailings dam* (PEER Report No. 2019/01).
Petley, D. (2020). The Luming Mine tailings accident in China, March 31, and more information on the Luming Mine tailings leak in China, April 1. *The Landslide Blog*. https://blogs.agu.org.landslideblog
Poole, A. (2008, December 4). *Tailings management at Omai Gold Mines—Catalyst for change*. https://bc-mlard.ca/files/presentations/2008-12-POOLE-tailing-mgt-omai-gold.pdf
Ramraj, R. (2001). *The Omai disaster in Guayana*. https://gammathetaupsilon.org/the-geographical-bulletin/2000s/volume43-2/article3.pdf
Reuters. (2019). Final settlement for Brazil's Samarco dam disaster could reach $19 bln, governor says. https://www.reuters.com/world/americas/exclusive-final-settlement-brazils-samarco-dam-disaster-could-reach-19-bln-2021-07-19/

Review Panel. (2015). *Mount Polley tailings storage facility breach.* Mount Polley Independent Expert Engineering Investigation and Review Panel. https://www.mountpolleyreviewpanel.ca/final-report

Rodríguez-Pacheco, R. L. (2019). Static liquefaction in tailings dam and flow failure. https://www.researchgate.net/publication/331277260

Simeoni, L., et al. (2017). The Stava catastrophic failure of July 19, 1985 (Italy): Technical-scientific data and socioeconomic aspects. *City Safety Energy Journal Issue 1 Planning and Land Safety,* 1–14.

Sixth Tone. (2021). China punishes 63 people over Heilongjiang tailings pond leak. https://www.sixthtone.com/news/1007537

SMCRA. (1977). *Surface mining control and reclamation act*, US Congress, August 1977.

Souza, S. (2024, December 3). Vale commissions a tailings-free Capanema iron ore project. *E&MJ Engineering and Mining Journal, Breaking News.*

Stefanek, P., et al. (2017). Surface tailings disposal at the Zelazny Most TSF, today and into the future. In *Paste 2017*. https://papers.acg.uwa.edu.au/p/1752_24_Stejanek/

Terzaghi, K. (1929). The effects of minor geologic details on the safety of dams. *Transactions of the American Institute of Mining and Metallurgical Engineers (AIME), Technical Publication, 215*, 31–44.

Touileb, B. N. (2013, April). On the consideration of aftershocks in dam design and safety. In *9th Symposium of the European Club*, Venice, Italy.

Tuv Sud. (2017). Periodic review of dam safety.

TVA. (2009, July 23). *Review of the Kingston fossil plant ash spill root cause study and observations about ash management* (39 pp. and Appendices A to E). Tennessee Valley Authority Office of the Inspector General, Inspection Report. https://www.oversight.gov/sites/default/files/oig-reports/TVA/2008-12283-02.pdf

UN DHA. (1995, August 19–30). *Guayana cyanide mine waste spill information reports No. 1 to No. 4.*

USGS. (2000, May 12–19). *An overview of mining-related environmental and human health issues, Marinduque Island, Philippines: Observations from a joint U.S. Geological Survey—Armed Forces Institute of Pathology reconnaissance field evaluation* (46 pp.).

Vick, S.G. (1995, November 16). Preliminary report on technical causation—Omai tailings dam failure. *Dam Review Team.*

Vick, S. G. (1996, September). Failure of the Omai tailings dam. *Geotechnical News.*

Villavicencio, G., Espinace, R., Palma, J., Fourie, A., & Valenzuela, P. (2014). Failures of sand tailings dams in a highly seismic country. *Canadian Geotechnical Journal, 51*, 449–464.

Wagener, F. (1997). The Merriespruit slimes dam failure: Overview and lessons learnt. *SAICE Journal, 39*(3):11– 15.

Wahler and Associates. (1973). *Analysis of coal refuse dam failure—Middle Fork Buffalo Creek, Saunders, West Virginia* (2 Vols.). National Technical Information Service, US Department of Commerce.

West Virginia. (1973). *The Buffalo Creek flood and disaster*. Official Report from the Governor's Ad Hoc Commission of Inquiry, West Virginia Archives and History. http://www.wvculture.org/history/archivesindex.aspx

WWF. (2002). *Mining in Doñana—Learned lessons.* WWF Spain, http://awsassets.wwf.es/downloads/mining_in_donana_english.pdf

Zhu, F., Zhang, W., & Puzrin, A. M. (2024). The slip surface mechanism of delayed failure of the Brumadinho tailings dam in 2019. *Communications Earth & Environment*. https://doi.org/10.1038/s43247-023-01086-09

Zieliski, S., & Stefanek, P. (2021). Measures to reduce the environmental impact of the Żelazny Most tailings storage facility. *IOP Conference Series: Earth and Environmental Science, 684*, 012017.

3·28伊春鹿鸣矿业尾矿库泄漏事故 2020. https://baike.baidu.com/item/3%C2%B728%E4%BC%8A%E6%98%A5%E9%B9%BF%E9%B8%A3%E7%9F%BF%E4%B8%9A%E5%B0%BE%E7%9F%BF%E5%BA%93%E6%B3%84%E6%BC%8F%E4%BA%8B%E6%95%85/49875870

黑龙江伊春鹿鸣尾矿库矿砂泄漏点仍未完成封堵，专家称尾矿坝无安全风险 2020. http://www.jxbbs.cc/wap/37114-1.html

伊春鹿鸣矿业尾矿砂泄漏事故进展: 污水团前锋已进入呼兰河 2020. https://www.thepaper.cn/newsDetail_forward_6766736

Epilogue

This monograph depicts tailings dam failure occurrence in 21 different settings that highlights the inherent dilemma we are in regarding the sustainability of mining industry in the twenty-first century and beyond. On the one hand, mining industry provides vital mineral supply to a modern world. On the other hand, due to the historical socio-economic development to date the industry has not performed adequately in protecting the safety of people and environment impacted by its operation.

Following a series of major tailings dam failures in recent years, there appears to be a willingness in the mining industry to face up to the current unsettled and non-sustainable practice associated with the fragility of tailings dams. Such uneasiness appeared before. Back in the late 1990s a series of large dam incidents/failures also sprang up in different parts of the world. These mishaps raised concerns of the public and regulators in the jurisdictions involved. Moreover, United Nations, other governmental agencies, and international financial agencies supporting mining developments in developing countries also attempted to rectify the situation. Numerous workshops were held, and guidelines regarding the proper design and management of tailings facilities were issued. These efforts helped to tighten up tailings dam practice to some extent. Three decades or so later, we seem to have fallen into the same rut again.

As tailings dams are built by the mining industry to serve its need to store the vast amount of fine-grained waste by-products, it is fitting for us to focus on the industry's latest drive to set up a Global Industry Standard on Tailings Management. The purpose of this standard is aiming at an ultimate goal of zero harm to people and the environment in tailings operations around the world. The goal is both noble and so dearly needed by the public. Discussed in the following are the latest initiatives undertaken by the mining industry, lessons gleaned from this monograph, and other useful information related to the safe operation of tailings storage facilities.

Global Industry Standard on Tailings Management (GISTM) Initiative (2018–2025)

The GISTM initiative started with the proposal by Professor Morgenstern (2018) ended with the founding of the Global Tailings Management Institute (GTMI, 2025).

R. C. Lo, *Tailings Dam Failure*, https://doi.org/10.1007/978-3-030-87738-5

The Institute will oversee the implementation of, and conformance with, the Global Industry Standard on Tailings Management (GISTM). Companies seeking to meet the Standard are required to prioritise safety, and enhance accountability through public disclosures at all tailings facilities. The GTMI will be responsible for managing the assurance framework through which tailings facilities will be independently audited and certified against the GISTM, by qualified third-party assessors. This initiative represents the latest drive of the global communities to face up the dilemma in the long evolution of tailings dam management practice. We are encouraged by the establishment of the Global Tailings Management Institute, whose mission is to provide a global framework for safety-focused assurance program to enhance the accountability of tailings-facility operators.

Professor Morgenstern (2018) proposed for the International Council of Mining and Metal to support a system of Performance-Based Risk-Informed Design and Safety Review and to facilitate its adoption in practice through the life cycle of design, construction, operation, and closure of tailings storage facilities. He pointed out that the system should be constrained by reliable evidence to a greater degree than is currently the case. Subsequently, UNEP (2020) published a Global Industry Standard on Tailings Management (GISTM). The Standard requires tailings operators to take responsibility and prioritize the safety of tailings facilities, through all phases of a facility's lifecycle, including closure and post-closure. It further requires the disclosure of relevant information to support public accountability. The document includes the Preamble, fifteen Principles and associated Requirements, the Glossary, and Annexes. The Standard is to be supported by implementation protocols that will provide detailed guidance for certification and equivalence with other standards. The 15 Principles are listed below:

1. Respect the rights of project-affected people and meaningfully engage them at all phases of the tailing facility lifecycle, including closure;
2. Develop and maintain an interdisciplinary knowledge base to support safe tailings management throughout the tailings facility lifecycle, including closure;
3. Use all elements of the knowledge base—social, environmental, local economic, and technical—to inform decisions throughout the tailings facility lifecycle, including closure;
4. Develop plans and design criteria for the tailings facility to minimize risk for all phases of its lifecycle, including closure and post-closure;
5. Develop a robust design that integrates the knowledge base and minimizes the risk of failure to people and the environment for all phases of the tailings facility lifecycle, including closure and post-closure;
6. Plan, build, and operate the tailings facility to manage risk at all phases of the tailings facility lifecycle, including closure and post-closure;
7. Design, implement, and operate monitoring systems to manage risk at all phases of the facility lifecycle, including closure;
8. Establish policies, systems, and accountabilities to support the safety and integrity of the tailings facility;
9. Appoint and empower an Engineer of Record;

10. Establish and implement levels of review as part of strong quality and risk management system for all phases of the tailings facility lifecycle, including closure;
11. Develop an organizational culture that promotes learning, communication, and early problem recognition;
12. Establish a process for reporting and addressing concerns and implement whistleblower protections;
13. Prepare for an emergency response to tailings facility failures;
14. Prepare for long-term recovery in the event of catastrophic failure; and
15. Publicly disclose and provide access to information about the tailings facility to support public accountability.

ICMM (2020) provided Equivalency Benchmarks with the following organizations on Mining Principles:

1. Aluminium Stewardship Initiatives;
2. The Copper Mark;
3. Responsible Minerals Initiative;
4. World Gold Council;
5. Towards Sustainable Mining; and
6. Risk Readiness Assessment.

Global Tailings Management Institute (GTMI, 2025) was established on January 21, 2025 in South Africa by ICMM, UNEP and PRI (Principles for Responsible Investment). The goal is to oversee an independent assessment process through which tailings facilities will be audited and certified. The Institute will oversee the implementation of, and conformance with, the Global Industry Standard on Tailings Management (GISTM). Companies seeking to meet the Standard are required to prioritise safety, and enhance accountability through public disclosures at all tailings facilities. The GTMI will be responsible for managing the assurance framework through which tailings facilities will be independently audited and certified against the GISTM, by qualified third-party assessors.

Lessons Gleaned from the Monograph

Key lessons gleaned from the monograph are highlighted in the following five aspects:

Related to Tailings Chemical Characteristics

The storage of mine tailings introduces a potential environmental degradation issue related to acid-generation and/or metal-leaching geochemical reactions. These reactions are important in our understanding of how they might degrade the quality of surface and groundwater in a tailings impoundment and further spread the degradation to the surrounding areas. Since 1998 INAP (inap.com.au), the International Network for Acid Prevention, has been actively involved in the development and sharing of a knowledge base covering collective experience from mining companies around the world in dealing with the acid and metalliferous drainage problem.

McLeod and Murray (2003) discussed some of the differences relative to the design of water versus tailings dams.

The water quality issue often subtly affects tailings dam safety. Because the acid and metalliferous drainage problem is a ubiquitous issue in the majority of tailings impoundments, maintenance of water quality of any discharge from a mining property in compliance with regulatory criteria is an ongoing challenge for an operating mine. Moreover, the issue often leads to a bone of contention between the mining operator and its regulatory agencies. This continual tussle tends to consume the time and effort of the mine and its regulators and often distracts both parties' attention from the critical dam safety issue of losing the precious impoundment capacity to safely store water to prevent dam overtopping failure. The longer the impasse continues, the odds of dam overtopping failure increase. As illustrated in many case histories in Chap. 3, one could appreciate many examples of such danger.

Related to Upstream Tailings Dam

Villavicencio et al. (2014) provided an excellent overview of the evolution of Chilean sand tailings dams engineering practice and regulatory framework including a summary of near a century of dam failures from 1915 to 2010.

As explained in Sect. 1.3, an upstream tailings dam generally has low seismic resistance, and could even fail under static gravity loading due to the phenomenon of spontaneous liquefaction. The recent rash of upstream tailings dam flow failures by static liquefaction, covered in Chap. 3 of case histories, reminded us that it is time to scrutinize all upstream tailings dams, especially those with a significant portion of the dam being saturated. We have learned from a rare pair of monitoring videos as well as eyewitnesses' accounts that the progress of static liquefaction failure of a tailings dam can happen at an amazing speed. The 83 m high Feijão Tailings Dam I in Brazil collapsed in about 6 min from a normal-looking solid embankment dam to a completely disintegrated tailings mudflow. The larger 100 m high Fundão Tailings Dam also in Brazil failed suddenly in a similar fashion. The forensic investigations of these two recent failures once again reminded us of the fragility of upstream tailings dams and the severity of the consequences of their failures.

Related to Regulatory Role

Governments in developing countries often lack adequate institutional capacity in dealing with the mining industry. Even in developed countries, governments could find themselves caught in a balancing act between promoting the growth of the mining industry on one hand and regulating mining projects to comply with the socio-economic-environmental development criteria on the other hand. It is a difficult role fraught with conflicting demands. However, maintenance of tailings dam safety relies on regulatory bodies fulfill their mandates diligently and effectively.

Related to Emergency Response

The emergency of a tailings impoundment develops when its safety is threatened. It could be threatened by the deterioration of stability of the tailings dam or its adjacent natural slope, internal erosion (piping) of a dam or its abutment/foundation, earthquake, the rise of supernatant pond level, or prolonged and/or heavy precipitation. The emergency could develop gradually or suddenly. It needs to be identified by monitoring the performance of the dam and impoundment visually and using instrumentation. The recent development of remote sensing monitoring of dam performance offers efficiency and is worthy of wider application (ASCE, 2021).

To prevent being caught off guard, the emergency response should be thoroughly planned, updated, and improved by regular reviews.

Related to Dam-Safety Stakeholders

Among all stakeholders, the group on the receiving end of inundation or mud-burying in an event of a dam breach is often the least capable of protecting themselves. Their precarious lot needs to be improved by the government which seems to be the only stakeholder that has both the power and responsibility to protect the potential victims of a tailings disaster.

Engineering Aspects of Tailings Dam Safety

Tailings dams are important, large hydraulic structures serving a vital industrial function of retaining vast amount of fine-grained tailings mine waste for the public. Due to their vulnerability to incidents and failures with catastrophic consequences, their safety and integrity need the sustained maintenance effort by all stakeholders. International Commission on Large Dams (ICOLD) with its world leadership in the design, construction, maintenance, and operation of large dams has an important role to play. Its state-of-the-art technical bulletins on embankment/tailings dam, natural and human hazards/factors, and incidents/failures are summarized in Table A.1 for easy reference to the specific subject or problem at hand. In addition, ICOLD's affiliated national commissions also have relevant references appropriate to the specific nation.

Future of Tailings Facility

It takes knowledge and vigilance to pursue the safety objective in a virtuous cycle of continual monitoring, review, and improvement. The recent dam incidents and failures did give a jolt to the dam community, and raise a crisis of public trust in dam safety. Meaningful actions are required for the mining industry to face up this crisis. Following thoughts are offered with the hope that they could help transform the current crisis into a renewed driving force compelling the industry to reduce the frequency as well as the severity of future dam incidents:

- Dam safety needs to have a much higher priority so that it would not be lost in the mind of all stakeholders over the long term when other more urgent matters tend to grab the operators' attention and weaken their safety-priority conviction. It is expected that more resources such as time and funds would be required to implement safety priority.
- All stakeholders, especially those potential victims of dam disasters, should be able to participate in the ongoing monitoring of dam performance as well as

Table A.1 Embankment/tailings dam related subjects and relevant ICOLD bulletins

Subjects	ICOLD bulletin number and abbreviated title
Embankment/tailings dam features	172-Spillway, 157-Small Dams, 151-Residual Soils, 150-Cutoffs, 141-Rockfill Dams, 135-Geomembrane Sealings, 134-Weak Rocks, 129-Foundations, 126-Roller-Compacted Dams, 123-Appurtenant Structures, 114-Bituminous Facing, 106-Tailings Dams, 95-Granular Filters/Drains, 92-Rockfill Materials, 91-Upstream Slope Protection, 89-Reinforced Rockfill and Reinforced Fill, 88-Rock Foundations, 81-Spillways and Air Entrainment
Hazards/factors affecting dam safety	170-Flood and Dam Safety, 169-Climate Change, 168-Operation, Maintenance and Rehabilitation, 167-Regulation of Dam Safety, 166-Post-earthquake Inspection, 164-Internal Erosion, 162-Hydraulic Processes, 160-Decommissioning, 159-Dams and Environment Supplement, 158-Surveillance, 156-Flood Risk, 154-Safety Management, 153-Tailings Dams Post-closure Performance, 148-Seismic Parameters, 146-Re-settlement, 143-Ancient Dams, 142-Extreme Floods, 139-Tailings Dam Safety, 137-Reservoirs and Seismicity, 133-Permafrost, 131-Flood Mitigation, 130-Risk Assessment, 125-Floods, 124-Landslides, 120-Seismic Design, 119-Rehabilitation, 118-Monitoring, 114-Automated Monitoring, 113-Seismic Observation, 112-Neotectonics, 105-Cold Climate, 104-Tailings Dams Monitoring, 103-Tailings Dams and Environment, 102-Hydraulic Equipment Vibrations, 101-Tailings Dams Operation, 98-Tailings Dams and Seismicity, 97-Tailings Dams Drainage, 93-Ageing, 90-Dams and Environment, 194-Tailings Dam Safety
Incidents/failures	121-Tailings Dams Risk, 99-Dam Failures

decision-making related to dam safety issues. An appropriate mechanism needs to be set up with the leadership of the government to establish this practice. The current situation that potential victims, whose life and livelihood could be ruined in a dam failure, do not even have a say in these important matters is no longer acceptable. It is both unconscionable and unsustainable.

- One positive development occurred after the Brumadinho (upstream) tailings dam failure. The State of Minas Gerais, Brazil banned upstream tailings dams. Banning upstream tailings dams would eliminate a type of tailings dam fraught with pitfalls. The stake is high to continue the indiscrete adoption of this inherently vulnerable type of tailings dam. The prudent approach should be leaning toward a robust, conservative design. The onus should be on those who propose any design that appears vulnerable to prove that the design is indeed safe for the proposed project because of specially favourable site and dam conditions or observable affirmative evidence of safe dam performance.
- Das et al. (2024) proposed an objective, probabilistic-learning algorithm to detect and characterize the risk of a slope failure, based on spectral analysis of serially correlated displacement time-series data. The results of their study demonstrate that the combination of spectral methods and second order statistical properties of InSAR displacement time-series data can reveal signs of a transition into

an unstable deformation regime, and that this algorithm can provide sufficient early-warning that could help arrest the process of development of a potentially catastrophic dam slope failure.

- United Nations Economic Commission for Europe (UNECE) partnered with Value.Space to use satellites to assess the risks of mine tailings storage facilities (Lasley, 2024). UNECE and Value.Space planned to monitor up to 10 tailings storage facilities. The intention is to analyze proactively the risks and reveals potential structural issues that can be addressed before a major failure occurs. As InSAR data analysis was successfully applied in forensic investigations of tailings dam failures, it is encouraging to see UNECE and Value.Space embark on this important initiative to monitor the actual performance of existing tailings storage facilities. It is hoped endeavours like this initiative will provide reliable and effective monitoring regimes for detecting early warning signs of potential threats to the safety of existing tailings dams.
- Vale recently commissioned the Capanema Maximization Project, near Itabirito in Minas Gerais (see Souza, 2024). The site is part of Vale's Operational Complex in Mariana and is expected to increase iron ore production by approximately 15 million metric tons per year towards the 2026 production guidance of 340–360 million mt. Vale is to restart the Capanema project using dry processing and eliminating the use of tailings dams for tailings storage. Moreover, the project will produce sinter feed at low cost and will use autonomous trucks, which will improve safety and reduce emissions of greenhouse gas CO_2.

List of Further Readings

ASCE. (2021). Remote sensing for monitoring embankments, dams, and slopes—Recent advances. In *Embankments, Dams, and Slopes Technical Committee, GSP 322* (102 pp.).

Das, S., Priyadarshana, A., & Grebby, S. (2024). Monitoring the risk of a tailings dam collapse through spectral analysis of satellite InSAR time-series data. *Stochastic Environmental Research and Risk Assessment, 38*, 2911–2926, https://doi.org/10.1007/s00477-024-02713-3

GTMI. (2025). *GTMI launch press release.* https://thegtmi.org/wp-content/uploads/2025/01/GTMI-launch-press-release-final-Updated.pdf

ICMM. (2020, December). *Mining principles: Equivalency benchmarks.* International Council on Mining and Metals.

ICOLD. (2024). *Tailings dam safety* (ICOLD bulletin 194 draft).

Lasley, S. (2024, December 2). UN to monitor mining facilities from space. *Metal Tech News.*

McLeod, H., & Murray, L. (2003, June 15). Tailings dam versus a water dam—What is the design difference? In *Proceedings of ICOLD Symposium on Major Challenges in Tailings Dams.*

Morgenstern, N. R. (2018). Geotechnical risk, regulation, and public policy. Sixth Victor de Mello Lecture. *Soils and Rocks, São Paulo, 41*(2), 107–129.

Souza, S. (2024, December 3). Vale commissions a tailings-free Capanema iron ore project. *E&MJ Engineering and Mining Journal, Breaking News.*

UNEP. (2020, August). *Global industry standard on tailings management* (40 pp.). United Nations Environment Program, International Council on Mining and Metals, and Principles for Responsible Investment.

Villavicentio, G., Espinace, R., Palma, J., Fourie, A., & Valenzuela, P. (2014). Failures of sand tailings dams in a highly seismic country. *Canadian Geotechnical Journal, 51*, 449–464.

Glossary of Technical Terms

Centreline dam Centreline tailings dam could be considered as an adaptation of downstream and upstream dam methods to meet the idiosyncrasy of tailings storage requirement. The central and downstream portions of the centreline dam can be engineered to be compatible with the natural setting of the dam site and project safety and economic requirements. Thus, the centreline dam can be as safe as the downstream dam. However, it is substantially less costly than the downstream dam. The reason for the cost-effectiveness of the centreline dam lies in the fact that its upstream stability relies on the weight rather than the shear strength of tailings. Therefore, the upstream portion of the centreline dam can adopt the relatively cheaper hydraulic-fill construction method used in the upstream dam.

Concrete-faced rockfill dam (CFRD) Concrete-faced rockfill dam is a rockfill embankment dam with an impervious reinforced-concrete slab facing. A CFRD also has a plinth at the base of the upstream concrete slab facing, from which a concrete seepage cutoff wall could be installed in the dam foundation, as required.

Dam failure versus dam incident Dam failure refers to cases involving a dam breach with the ensuing release of stored fluid and/or solid content, while dam incident refers to cases of relatively less damaging consequences where no dam breach occurs.

Downstream dam A downstream dam is directly adopted from a conventional earth- and/or rockfill water-retention embankment dam for tailings storage. It does not, however, take advantage of the fact that tailings is much heavier than water, and the duration of tailings dam construction is significantly longer than that for a water dam. Consequently, the cost of the downstream tailings dam is higher than both the centreline dam and upstream dam construction methods.

Effective stress, neutral stress and total stress Effective stress is an important concept in soil mechanics as elucidated by Dr. Karl Terzaghi about a century ago. Soil is a three-phase material made of soil grains, void water and void air. When soil is stressed, only that portion transmitted by soil grains contributed to its shear strength, thus called effective stress. Those portions transmitted by void water and/or void air are called neutral stress, as these stresses do not contribute to soil shear strength. The sum of effective and neutral stresses is total stress.

R. C. Lo, *Tailings Dam Failure*, https://doi.org/10.1007/978-3-030-87738-5

Embankment dam Embankment dam is a versatile type of dam that can be founded on a variety of soil and rock sites. In general, an embankment dam is an earth- and/or rockfill retaining structure, including a concrete-faced rockfill structure, which can be used to store water and/or a mixture of fluid/solid.

Evacuation order A critically important order issued by the government agency that has jurisdiction over the emergency evacuation of a community threatened by inundation flood and/or buried by debris flow resulting from a potential breach of a tailings dam. The mining company involved in such a scenario is obligated to report to the government agency promptly to allow sufficient time for the emergency responders to organize and carry out safely and effectively the evacuation operation.

Hydrocyclone (cyclone) Is a hydraulic device with no moving parts that makes use of the centrifugal principle to separate materials of different sizes. Cyclones are used to separate coarse and fine fractions of total tailings. The coarse tailings are used for constructing a tailings dam, while the fine tailings are stored within the tailings impoundment enclosed by the tailings dam.

Liquefaction For strain-softening soil, as the shear stress–strain curve reaches its peak strength, under sustained undrained shear-loading, the soil could fail suddenly and continue to deform and flow as if it has been transformed to a viscous material, or "liquified".

Observational approach The observational method is developed by Drs. Karl Terzaghi and Ralph Peck as they coauthored a textbook on Soil Mechanics in Engineering Practice. The essence of the method is experimental, i.e. "learn-as-you-go". Beginning with a most probable working hypothesis based on factual information acquired by subsoil investigation. Followed by applying soil mechanics to arrive at a design and selecting the most critical parameters for ongoing observation, such as pore pressure or displacement. As construction proceeds, measured values of observed parameters are compared with those calculated based on the original working hypothesis to test the working hypothesis. Thus, the working hypothesis is constantly being confirmed or improved by making appropriate adjustments until the construction is successfully completed.

Piping failure Internal erosion of soil fine particles from one embankment zone to its adjacent zone by seepage force due to the incompatibility of the grain-size distributions of the two adjacent soils at the boundary. If this erosion process is allowed to continue, the erosion "pipe" could extend from the downstream end to upstream and eventually daylight to the supernatant pond and lead to a dam breach.

Spontaneous liquefaction Liquefaction of soil could be triggered under static or dynamic loading. Under static loading, the process could appear spontaneously as if no additional external force is needed to trigger it, while dynamic liquefaction is triggered by external seismic or blasting dynamic forces.

Starter dam Is the initial portion of a tailings dam that is constructed to provide the initial storage volume for tailings as well as water required during the start-up phase of a mining operation. Subsequently, the tailings dam is raised in stages to meet the storage requirement of ongoing tailings production.

Static liquefaction Static liquefaction is a special type of liquefaction phenomenon which occurs in materials that are inherently unstable due to the physico-chemical circumstances accompanied to the history of their deposition and subsequent existence. For tailings storage facilities, the phenomenon of static liquefaction is well known to be associated with tailings dams constructed by the upstream-dam-method as described in Sect. 1.4 Tailings Dam. Unfortunately, due to the cost advantage of the upstream dam construction method offered to mining companies, the method has been continually used to date with catastrophic consequences as shown in Chap. 3—Representative Case Histories and Lessons Learned. With the recent movement associated with Global Industry Standard on Tailings Management, we could hope that a pivotal milestone will be established soon. Essentially, as most mining companies adopt the Global Industry Standard, the upstream-dam-construction method, with its attendant risk of static liquefaction leading to tailings dam flow failure, will be eliminated.

Supernatant pond Supernatant pond refers to the clarified water body above the settled tailings stored in a tailings impoundment.

Tailings Tailings, sometimes referred to as fine-grained waste solid particles, are generated in the process of winning minerals from the ore. Because the quantity of tailings produced by a modern mine is enormous, the safety of tailings storage facilities becomes an important public safety and environmental protection issue.

Tailings slurry density (pulp density) The ratio expressed as a percentage of the weight of solid divided by the combined weights of both solid and liquid in a tailings slurry. Slurry density has significant influences on tailings transportation, deposition and settlement processes.

Upstream dam A tailings dam constructed by the upstream method has only a small portion of the dam, can be constructed as engineered dam sections at the downstream toe as the Starter Dam and along the downstream slope as a series of small raised dams used to raise the crest of the tailings dam in small increments above the initial Starter Dam crest. The bulk of the dam is built by tailings transported to the dam site by hydraulic means. Thus, the physical and engineering properties of the main body of the tailings dam is governed by the deposition and settlement of tailings in braided streams of tailings slurry flow inside the tailings impoundment. Since this process is not subject to engineering control, the density and strength of the bulk damfill is practically unknown. In general, upstream dam is least costly to construct, and, unsurprisingly, has the poorest safety record.

FSC
www.fsc.org
MIX
Papier aus verantwortungsvollen Quellen
Paper from responsible sources
FSC® C105338

If you have any concerns about our products,
you can contact us on
ProductSafety@springernature.com

In case Publisher is established outside the EU,
the EU authorized representative is:
Springer Nature Customer Service Center GmbH
Europaplatz 3, 69115 Heidelberg, Germany

Printed by Libri Plureos GmbH
in Hamburg, Germany